T0229655

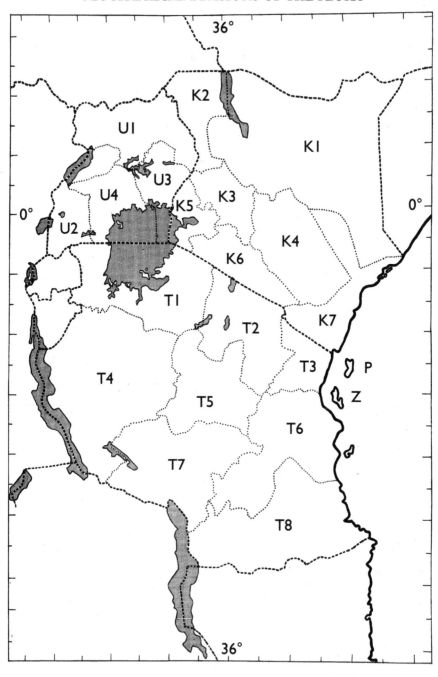

FLORA OF TROPICAL EAST AFRICA

APOCYNACEAE (Part 1)

E.A. OMINO*

Trees, shrubs, lianas, or sometimes herbs (not in our area, except introduced ones); latex nearly always present, usually white, sometimes clear, yellow or red. Leaves simple, opposite, whorled, or less often alternate (as in *Adenium*), pinnately veined, entire, rarely stipulate. Flowers bisexual, mostly actinomorphic, 5- or rarely 4-merous, fragrant. Calyx often with colleters inside, free or united at the base, imbricate in bud. Corolla tubular, contorted or occasionally valvate, free, overlapping to the right or left in bud, sometimes with a corona. Stamens included or exserted, epipetalous, free or connate to the style; filaments often very short, epipetalous; anthers frequently triangular, of two cells longitudinally dehiscent, often partly sterile, sometimes with apical appendages. Ovary superior or sometimes partly inferior, 1-celled and with 2 parietal placentas, 2-celled and with an axile placenta in each cell or composed of 2, or more, separate or at the base partly united carpels each with an adaxial placenta; ovules 2 to many; style one, often split at the base when carpels more or less separate; pistil head composed of a large variously shaped stigmatic basal part and a stigmoid apex; disk annular, cupular, composed of separate glands, or absent. Fruit entire or consisting of two (rarely more) separate or partly united carpels, baccate, drupaceous, or follicular. Seeds in dry fruits often winged or with a coma, mostly with an endosperm and a large embryo.

A family of about 180 genera and about 1700 species, mainly in the tropics.

Recently the family Asclepiadaceae has been subsumed into the Apocynaceae: see M.E. Endress & P.V. Bruyns in Botanical Review 66: 56 (2000). A key to the subfamilies based on that given by Endress & Bruyns is included below.

KEY TO APOCYNACEAE SENSU LATO

1. Anthers adnate to style head; corolla lobes usually overlapping to the right or valvate, rarely overlapping to the left; fruit dehiscent, almost always apocarpous, a pair of follicles, sometimes reduced to one; seeds small, compressed, usually with a tuft of hairs · 2

   Anthers free from style head; corolla lobes usually overlapping to the left; fruit dehiscent or indehiscent, syncarpous or apocarpous, a berry, drupe, follicle or capsule; seeds usually without tuft of hairs, often with wings or arils · · · · · · · ·   subfamily Rauvolfioideae (genera 1–18)

* The author would like to honour Dr A.J.M. Leeuwenberg, who has published extensively on this family; and who has trained many taxonomists, the present author amongst them.
  A. Katende has kindly provided much information on uses of Ugandan Apocynaceae.
* Address of author: c/o. PROTA, ICRAF, P.O. Box 30677, Nairobi, Kenya.

1

2. Nectaries, if present, in a ring around the base of ovary; anthers 4-locular; pollen usually shed as monads; style head secretions for pollen transport undifferentiated, foamy or gummy · · · · · · · · · · · · · · · ·   subfamily Apocynoideae (genera 19–30)

Nectaries in troughs on staminal bases or staminal tube; anthers 2–4-locular; pollen shed as tetrads or in pollinia; style head secretions forming differentiated translators with a sticky end or consisting of a clip with two flexible arms · · · · · · ·   subfamilies Asclepiadoideae, Periplocoideae, Secamonoideae (in part 2)

KEY TO SPECIES OF APOCYNACEAE SENSU STRICTO
(SUBFAMILIES RAUVOLFIOIDEAE AND APOCYNOIDEAE)

1. Leaves opposite or verticillate; plant not succulent · 2
   Leaves alternate or in bunches set close together, confined to apices of branchlets · 52
2. Leaves in whorls of 3–9 · 3
   Leaves opposite · 6
3. Climbers or shrubs; cultivated · · · · · · · · ·   *Allamanda* (p. 6)
   Trees or shrubs · 4
4. Inflorescence axillary, fasciculate, (sub)sessile and usually ramiflorous, up to 40-flowered   10. **Pleiocarpa** (p. 39)
   Inflorescence terminal (rarely axillary), branched, pedunculate, congested or lax · · · · · · · · · · · · · · · · · · · 5
5. Leaves in whorls of 4–9; secondary veins straight, forming an angle of 80–90° with the midrib and a submarginal vein; fruits linear, follicular; seeds with a coma at both ends ·   17. **Alstonia** (p. 58)
   Leaves in whorls of 3–6; secondary veins gradually curved towards the margin, forming an angle of 60–70° with the midrib; fruits drupaceous, globose to ellipsoid; seeds without a coma · · · · · · · ·   18. **Rauvolfia** (p. 60)
   Leaves strictly in threes, or some opposite; secondary veins very many, only slightly curved over the small width of the narrow blade; fruits linear, follicular; (cultivated)   *Nerium* (p. 10)
6. Plants spiny ·   1. **Carissa** (p. 11)
   Plants unarmed · 7
7. Herbs · 8
   Plants woody · 9
8. Erect herb with reddish-pink or white flowers   *Catharanthus* (p. 6)
   Creeping herb with rich blue flowers · · · · ·   *Vinca* (p. 10)
9. Lianas or climbers · 10
   Trees or shrubs · 20

10. Tendrils present at the forks of branches; corolla lobes overlapping to the left; fruits globose 11

Tendrils absent; corolla lobes overlapping to the right; fruits oblong 15

11. Inflorescence of elongate terminal panicles 12

Inflorescence contracted, short and often clustered, terminal or axillary, exceptionally some of them elongate 13

12. Leaf base rounded or cuneate; stipules absent; ovary and fruit pubescent .... 6. **Ancylobotrys** (p. 28)

Leaf base subcordate, stipules interpetiolar, early caducous; ovary and fruit glabrous · 5. **Dictyophleba** (p. 26)

13. Only terminal inflorescence developed, dense, contracted; leaves rounded to obtuse at the apex (rarely acuminate); corolla lobes 19–41 mm long .... 7. **Saba** (p. 32)

Inflorescence terminal and/or axillary; leaves often acuminate; corolla lobes 1–17.5 mm long 14

14. Inflorescence axillary; anthers up to 0.5 mm long, hanging or pointing downwards; pistil never touching the stamens .... 4. **Clitandra** (p. 24)

Inflorescence terminal and/or axillary; anthers more than 0.8 mm long; pistil touching the stamens .... 3. **Landolphia** (p. 18)

15. Corolla with paired corona lobes; seed with long stalked hair plume .... 24. **Strophanthus** (p. 75)

Corolla with either single corona lobes or without corona; seed with hair plume but this not stalked 16

16. Cultivated climber; flowers > 10 cm long, white .... *Beaumontia* (p. 6)

Wild plants; flowers (much) smaller 17

17. Leaf petiole with glands on the adaxial side 18

Leaf petiole without glands .... 28. **Alafia** (p. 93)

18. Only terminal inflorescence developed; fruits winged or ridged .... 29. **Motandra** (p. 99)

Inflorescence both axillary and terminal; fruits neither winged nor ridged 19

19. Corolla with 5 alternipetalous corona scales at the mouth .... 31. **Oncinotis** (p. 105)

Corolla without corona scales at the mouth · 30. **Baissea** (p. 101)

20. Leaf apex with a hard mucro .... 2. **Acokanthera** (p. 14)

Leaf apex without a mucro 21

21. Key to plants with flowers present 22

Key to plants with fruit present 37

22. Corolla lobes overlapping to the left 23

Corolla lobes overlapping to the right 33

23. Corolla with corona or appendages at the mouth 24

Corolla without corona or appendages at the mouth (though in *Diplorhynchus* a small scale < 1 mm long is present between the bases of the lobes) 26

24. Corolla mouth with 20 filiform branched
    appendages · · · · · · · · · · · · · · · · · · · 22. **Pleioceras** (p. 71)
    Corolla with corona · 25
25. Corona undulate, 3 mm long; stamens
    included · 23. **Stephanostema** (p. 73)
    Corona 1–1.5 mm long; stamens exserted · 21. **Wrightia** (p. 69)
26. Corolla tube twisted, and often lobes twisted
    as well · 27
    Corolla tube and lobes straight (or only
    lobes occasionally twisted in *Hunteria*) · 28
27. Sepals united for two thirds or more · · · · · 11. **Voacanga** (p. 41)
    Sepals almost free · · · · · · · · · · · · · · · · · · 12. **Tabernaemontana** (p. 45)
28. Inflorescence in pairs at the forks of
    branches; corolla campanulate; leaves
    with colleters in the axils · · · · · · · · · · · 14. **Carvalhoa** (p. 52)
    Inflorescence terminal or axillary; corolla
    tubular or salver-shaped; leaves without
    colleters (except in *Callichilia*) · 29
29. Corolla lobes pilose at the base, with a
    glabrous scale between the bases of all
    lobes; fruits oblong · · · · · · · · · · · · · · · 16. **Diplorhynchus** (p. 56)
    Corolla lobes glabrous and without a scale in
    between; fruits ellipsoid, obovoid or
    globose · 30
30. Sepals glabrous on both sides; inflorescence
    predominantly axillary, ramiflorous · · · · · 10. **Pleiocarpa** (p. 39)
    Sepals glabrous outside, with colleters within;
    inflorescence predominantly terminal · 31
31. Sepals < 3 mm long · · · · · · · · · · · · · · · · · 9. **Hunteria** (p. 36)
    Sepals > 3 mm long · 32
32. Shrub 1–3 m high; leaves 4–12 × 1–4 cm;
    coastal Tanzania · · · · · · · · · · · · · · · · · 13. **Callichilia** (p. 50)
    Shrub or tree 4–35 m high; leaves 10–27 ×
    2–13 cm; Uganda · · · · · · · · · · · · · · · · · 8. **Picralima** (p. 34)
33. Corolla with 4–8 cm long tails · · · · · · · · · 24. **Strophanthus** (p. 75)
    Corolla lobes not tailed · 34
34. Petiole with glands near the base; sepals
    valvate · 27. **Holarrhena** (p. 92)
    Petiole without glands (though *Funtumia*
    and *Schizozygia* have glands in leaf axils);
    sepals imbricate · 35
35. Leaf petiole connate into a short ochrea;
    inflorescence terminal or axillary · 36
    Leaf petiole without ochrea; inflorescence in
    pairs at the forks of branches · · · · · · · · · 15. **Schizozygia** (p. 54)
36. Leaves with domatia in vein axils; corolla
    lobes glabrous to puberulous · · · · · · · · · 25. **Funtumia** (p. 86)
    Leaf without domatia in vein axils; corolla
    lobes densely hairy · · · · · · · · · · · · · · · 26. **Mascarenhasia** (p. 89)
37. (**Fruits**) Fruit more than 3 × as long as wide,
    or obliquely oblong · 38
    Fruit (sub-)globose, ovoid or ellipsoid · 47
38. Fruits yellow or pale orange, indented
    around seeds when dry · · · · · · · · · · · · 14. **Carvalhoa** (p. 52)
    Fruits green, grey or brown · 39

54. Shrub or tree of coastal forest or thicket;
    leaves > 1 cm wide; fruit mericarps 11–22
    × 1–2 cm; **P** ·   19. **Cerbera** (p. 65)
    Cultivated tree; leaves < 1 cm wide; fruit
    globular ·   *Thevetia* (p. 10)
55. Calyx > 3 mm long; fruit of partly connate
    mericarps ·   13. **Callichilia** (p. 50)
    Calyx < 3 mm long; fruit of two free mericarps   9. **Hunteria** (p. 36)

Many Apocynaceae are of horticultural interest, and several are cultivated in East Africa; the common cultivated species are included in the key.

Allamanda have very spiny subglobose fruits.

Allamanda schottii *Pohl* – T.T.C.L.: 48 (1949); sometimes mistakenly called *A. neriifolia* Hook. A shrub from S America known as the 'Golden trumpet bush'; leaves like those of *Nerium*; flowers small, yellow. Kenya, Nairobi City Park, Apr. 1953, *Bally* 8861! Tanzania, Amani, Aug. 1929, *Greenway* 1699! & Oct. 1969, *Ruffo* 272! & Rungwe District: Kiganga Estate, Oct. 1971, *Perdue & Kibuwa* 11657!

Allamanda cathartica *L.* – U.O.P.Z.: 114, ill. (1949). A vigorous climber from tropical S America, known as the 'Golden trumpet vine'. Leaves whorled, glossy; flowers golden, to 8 cm across. Kenya, Nairobi, June 1958, *Bell* H163/58/1. Tanzania, Uzaramo District: Kisarawe, by Kimani stream, possibly naturalized in riverine forest, Feb. 1977, *Wingfield* 3770! Fig. 1.1 (p. 7).
There is a cultivar 'Hendersonii' with chrome yellow flowers: Kenya, Nairobi, May 1948, *Starzenski* in *Bally* 6213! Tanzania, Amani, July 1950, *Verdcourt* 287!

Allamanda violacea *Gardn. & Field.* – T.T.C.L.: 48 (1949); U.O.P.Z.: 114 (1949). The 'Purple Allamanda' from tropical S America is widely cultivated. It is a shrub or weak climber with hairy branches and leaves and reddish-purple flowers. Kenya, Kilifi District: Kikambala, Nov. 1975, *Leeuwenberg* 10788! & Kilifi Creek, near lodge, Aug. 1975, *Greenway* 15545!; Malindi District: Watamu, Dec. 1969, *Kimani* 191! Tanzania, Amani, Oct. 1933, *Greenway* 3671!; Dar es Salaam, Jan. 1974, *Ruffo* 1065!; Morogoro, Nov. 1955, *Semsei* 2392! A specimen collected at Pangani, Langoni, Nov. 1956, *Tanner* 3411! is possibly an escape.

Beaumontia grandiflora (*Roxb.*) *Wallich* – U.O.P.Z.: 145 (1949). A climber from the Himalayas, sometimes known as the 'Easter-lily vine'. Young branches and leaves rusty; leaves slightly obovate, acuminate, to 30 cm long. Flowers white, to 15 cm long. Nairobi Arboretum, June 1952, *Williams Sangai* 460!

Catharanthus roseus (*L.*) *G. Don* – F.W.T.A. ed. 2, 2: 68 (1963); Codd in F.S.A. 26: 268 (1963); Plaizier in F.Z. 7, 2: 454, t. 106 (1985) & in Meded. Landbouwhogeschool 81 (9): 3, t. 1, phot. 1 (1981); U.K.W.F. ed. 2: 172 (1994). Syn. *Lochnera rosea* L. – U.O.P.Z.: 333, ill. (1949); F.P.U.: 118, t. 65 (1962); Wild Flow. E. Afr.: 142, t. 675 (1987). 'Madagascar Periwinkle', originally from Madagascar but widely cultivated and sometimes naturalized. Woody herb to 60 cm. Leaves oblanceolate. Flowers reddish pink or white with crimson centre. Uganda, W Nile District: NW of Maracha rest camp, Aug. 1953, *Chancellor* 84!; Mengo District: Entebbe, Jan. 1931, *Snowden* 1941! Kenya, Nairobi: Muguga Estate, June 1962, *Gichuru* 12!; Tsavo National Park East, Apr. 1965, *Hucks* 95!; Kilifi District: Jilore Forest Station, Nov. 1972, *Spjut* 2631! Tanzania, Mwanza, Oct. 1949, *Windisch-Graetner* in *Bally* 7573!; Kilimanjaro, Marangu, Apr. 1961, *Machangu* 1!; Rufiji District: Mafia I., Aug. 1937, *Greenway* 5162!; Zanzibar, Ber-el-Res, July 1950, *J.L. Smith* 108! Fig. 1.2 (p. 7).

FIG. 1. *ALLAMANDA CATHARTICA* — **1**, habit, × ¹/₃. *CATHARANTHUS ROSEUS* — **2**, habit, × ¹/₂. *PLUMERIA RUBRA* — **3**, habit, × ¹/₃. 1 with reference to *Brummitt* 15410; 2 with reference to *Chancellor* 84; 3 with reference to *Lye* 23404. Drawn by J. Williamson.

Fig. 2. *NERIUM OLEANDER* — **1**, flowering shoot, × ¹/₂; **2**, flower, × 1. From Verdcourt &
Trump, Poisonous plants of East Africa, drawn by M.E. Church.

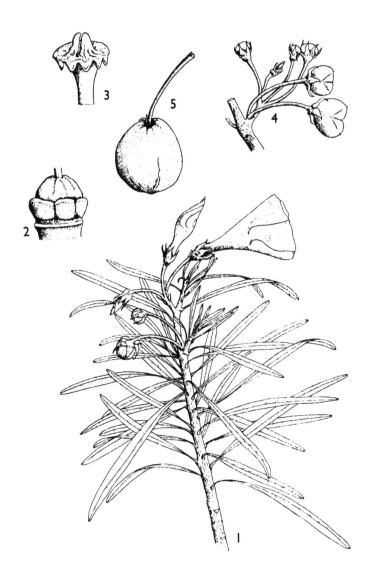

FIG. 3. *THEVETIA PERUVIANA* – **1**, flowering shoot, × ¹/₂; **2**, ovary and disk, × 5; **3**, apex of gynoecium, × 8; **4**, fruiting branchlet, × ¹/₂; **5**, mature fruit, × ¹/₂. From Verdcourt & Trump, Poisonous plants of East Africa, drawn by M.E. Church.

Mandevilla laxa (*Ruiz & Pavon*) *R.E. Woodson* (*M. suaveolens* (DC.) Lindl.) is a cultivated climber that has been collected in Kenya and Tanzania: Kenya, Naro Moru, Apr. 1977, *Patel & Patel* 188! Tanzania, Arusha, Dec. 1935, *Bennet & Montague* H83/35!

Mandevilla splendens (*Hook.*) *R.E. Woodson* or possibly **M. × amoena**. A cultivated climber with large rose-pink flowers. Kenya, Nairobi, Aug. 1965, *Stewart* in EA 13256! Tanzania, Lushoto District: Amani, Bustani, Dec. 1956, *Verdcourt* 1756!

Nerium oleander *L.* – T.T.C.L.: 53 (1949); U.O.P.Z.: 379 (1949); Verdcourt & Trump, Comm. Poison. Pl. E. Afr.: 131, t. 9 (1969). The oleander, from the Mediterranean and Asia; a shrub or small tree with opposite or ternate, thick and poisonous leaves, and attractive, fragrant flowers that are white, crimson or pink. Widely cultivated but only collected from Dar es Salaam Botanic Gardens, Nov. 1972, *Ruffo* 581! Fig. 2 (p. 8)

Plumeria obtusa *L.* Syn. *P. alba* sensu auctt. non L. – T.T.C.L.: 54 (1949). A cultivated shrub or tree from the Americas, to 8 m with white flowers. Zanzibar, Mbweni, Dec. 1930, *Greenway* 2675!

P. rubra *L.* – U.O.P.Z.: 417, ill. (1949). Syn. *P. acuminata* Ait. – T.T.C.L.: 54 (1949). 'Frangipani'; widely cultivated. Shrub or small tree from the Americas, to 9 m, much branched. Flowers white, yellow, pink or red. Uganda, Makerere, Feb. 1999, *Lye* 23404! & Busoga, Kaliro, Feb. 1999, *Lye et al.* 23462!. Kenya, Mombasa: Changamwe, Mar. 1902, *Kassner* 258b. Tanzania, Lushoto District: Amani, Aug. 1929, *Greenway* 1701! Fig. 1.3 (p. 7).

Thevetia peruviana (*Pers.*) *Merr.* – T.T.C.L.: 57 (1949); U.O.P.Z.: 470, ill. (1949); Verdcourt & Trump, Comm. Poison. Pl. E. Afr.: 141, t. 11 (1969). Sometimes known as 'Lucky Nut'. A fast-growing small tree from the Americas, widely cultivated and naturalized in relict forests along the lower Tana River, Kenya (Robertson, pers. comm.). Leaves almost linear, shiny. Flowers yellow (rarely pink), 3–6 cm long. Bark, roots, seed kernels and latex are highly poisonous. Kenya: Nairobi, Chiromo Estate, Nov. 1969, *Mathenge* 476!; Tsavo National Park, without date, *Sheldrick* TNP/E/54!; Kilifi District: Jilore Forest Station, Dec. 1969, *Perdue & Kibuwa* 10163! Tanzania: Tanga District: Korogwe, May 1966, *Semsei* 4009 & Tanga, Apr. 1979, *Grabner* 327!; Uzaramo District: Kisarawe, Aug. 1953, *Semsei* 1337! Fig. 3 (p. 9).

A second species of *Thevetia*, possibly *T. thevetioides* K. Schum., is common in Nairobi gardens; no herbarium specimens have been seen.

Trachelospermum jasminoides (*Lindl.*) *Lem.* A climber with white flowers from Asia, occasionally cultivated. Kenya. Trans Nzoia District: Kitale, T.H.E. Jackson's garden, Sep. 1959, *Verdcourt* 2448!; Nairobi, Chiromo Estate, Dec. 1970, *Mathenge* 727! & National Museums, Nov. 1982, *Mathenge* 82/173!

Vinca minor *L.* – U.K.W.F. ed. 2: 172 (1994). The 'Lesser Periwinkle' of Europe, sometimes naturalized in our area. Herb to 50 cm with long stoloniferous shoots. Leaves ovate. Flowers rich blue. Kenya, Kiambu District: Kinale Sawmills, Aug. 1964, *Gillett* 16155! Tanzania, Moshi District: Kilimanjaro, Londorossi Sawmill, Sep. 1993, *Grimshaw* 93/765!

1. CARISSA

L., Syst. Nat., ed.12, 2: 189 (1767) & Mant. Pl. Alt. 7: 52 (1767), nom. conserv.;
Kupicha in F.Z. 7, 2: 398–404 (1985); Leeuwenberg & van Dilst in Wageningen
Univ. Pap. 01.1: 1–109 (2001)

Arduina L., Syst. Nat., ed.12, 2: 136, 180 (1767)

Much branched shrubs, sometimes scandent, with simple or forked spines;
stipules absent. Leaves opposite, petiolate; blade ovate to suborbicular, with or
without a mucro at the apex, entire or slightly crenate at the margin. Inflorescence
a terminal corymb or cyme, dense to lax. Flowers 5- rarely 4-merous, fragrant. Sepals
imbricate. Corolla pink outside, white inside; tube cylindrical; lobes contorted,
overlapping to the left or right; stamens included, free from the style, inserted at the
middle or towards the apex of the corolla tube; ovary glabrous, 2-celled, with 1–4 or
rarely many ovules per locule; style terete; pistil head of two parts, a subglobose or
ellipsoid basal stigmatic part and a bilobed stigmoid apex. Fruit an ellipsoid or
ovoid berry with 1–8 seeds.

A genus of 6 species in continental Africa and ± 14 species in Asia, Australia, Madagascar, the
Comoros, Seychelles and Mascarenes.

C. carandas L. and *C. macrocarpa* (Eckl.) A.DC. [syn. *C. grandiflora* (E. Meyer) A.DC.], the Natal
plum, are sometimes cultivated (U.O.P.Z.: 173 (1949)). Kenya, Nairobi: Muthaiga, without date,
E. Polhill 225; Kwale District: Kwale Forest Station, Jan. 1970, *Perdue & Kibuwa* 10227.

1. Spines almost always simple; corolla lobes overlapping to the
   right; inflorescence terminal, dense · · · · · · · · · · · · · ·   2. *C. spinarum*
   Spines furcate or bifurcate, rarely simple; corolla lobes
   overlapping to the left; inflorescence at the fork of spines,
   few-flowered · 2
2. Secondary veins 12–18 or more; leaf margin crenate; flowers
   4-merous ·   3. *C. tetramera*
   Secondary veins 3–8; leaf margin entire; flowers 5-merous · · ·   1. *C. bispinosa*

1. **C. bispinosa** (*L.*) *Brenan* in Mem. N. Y. Bot. Gard. 8: 502 (1954); Codd in
F.S.A. 26: 255, t. 37/1 (1963); K.T.S.: 44 (1961); Kupicha in F.Z. 7, 2: 403 (1985);
K.T.S.L.: 478, map (1994); Leeuwenberg & van Dilst in Wageningen Univ. Pap.
01.1: 8, t. 1, 2, map (2001). Type: Mill. Ic. 2: t. 300 (1760), designated by Codd in
Bothalia 7: 450 (1961)

Shrub 0.5–6 m high, or tree to 10 m; spines thick, usually simple (in our area) or
furcate to bifurcate, 2.1–5 cm long. Leaves petiolate; petiole 1–4 mm long; blade
broadly elliptic, ovate or suborbicular, 1–11 cm long, 0.6–6 cm wide, obtuse or
mucronate at the apex, truncate at the base, glabrous; secondary veins obscure
below. Inflorescence terminal at the fork of spines, few-flowered. Flowers fragrant,
white with a pink tube; sepals ovate, 1.5–3 mm long, acuminate, minutely ciliate,
glabrous to pubescent; corolla tube 5–14 mm long, with long white hairs at the
mouth and base of lobes; lobes overlapping to the left, ovate, 3–6 mm long, 1–3 mm
wide, acuminate; stamens inserted near the top of the corolla tube. Fruit red, ovoid
or ellipsoid, 10–20 mm long, 4–10 mm in diameter, apiculate, glabrous, 1–2-seeded.

UGANDA. Kigezi District: between Nyakalembe and Mashogiri, *Gilbert Rogers & Gardner* 297
KENYA. Tana River District: Lango ya Simba, 6 Aug. 1988, *Robertson & Luke* 5367! Kilifi District:
   9 km on Bamba–Karimani road, 21. Nov. 1989, *Luke & Robertson* 2144! Kwale District: Gonja
   Forest Reserve, 18 Aug. 1993, *Luke & Luke* 3799B!
TANZANIA. Tanga District: Amboni, *Holst* 2474a

DISTR. **K** 7; Malawi, Mozambique, Zimbabwe, Botswana, Namibia, Swaziland, S Africa
HAB. Dry forest and coastal woodland; 0–400 m
USES. Fruit edible

SYN. *Arduina bispinosa* L., Mant. Pl. Alt.: 52 (1767)
   Carissa arduina (L.) Lam., Encycl. Méth. Bot. 1: 555 (1785), nom. illegit.

NOTE. The specimens cited from Kenya have much smaller leaves and the spines are simple, rarely furcate.

2. **C. spinarum** *L.*, Mantissa 1: 52 (1767) & Syst. ed. 12, 2: 189 (1770); Leeuwenberg & van Dilst in Wageningen Univ. Pap. 01.1: 35, t. 7–10, map 6–7 (2001). Type: India, without locality, *Koenig* s.n. (LINN 295.2, lecto., chosen by Huber in 1973)

Shrub 0.2–3 m, occasionally scrambling up to 20 m high; bark grey, slightly rough or smooth; spines simple, rarely forked, 0.5–6.2 cm long. Leaves petiolate; petiole 0.5–6 mm long; blade ovate, elliptic or almost orbicular, 1–8(–17) cm long, 0.5–6 cm wide, acute or rounded at the apex, with or without a hard mucro, cuneate or rounded at the base, glabrous or pubescent, with 3–5 pairs of secondary veins. Inflorescence terminal or occasionally axillary, dense-flowered cymes. Flowers fragrant, white inside, pink to red outside; sepals (narrowly) ovate, 1.5–4.5 mm long, 0.4–1.7 mm wide, acuminate, ciliate; corolla tube 5–25 mm long, tubular, widest at the throat, pubescent at the mouth and lower lobes; lobes overlapping to the right, obliquely oblong or ovate, 4–15 mm long, 1–7 mm wide, acute at the apex; stamens inserted near the apex of the tube. Fruit red to black, globose or ellipsoid, 5–25(–60) mm long, 3–20(–60) mm in diameter. Fig. 4 (p. 13).

UGANDA. Karamoja District: Kidepo National Park, 14 May 1972, *Synnott* 976!; Ankole District: Chahi Forest Reserve, 2 Feb. 1956, *Harker* 210!; Busoga District: Bunya, Nov. 1937, *Webb* 29!
KENYA. Northern Frontier District: Maralal, 13 Apr. 1975, *Fratkin* 37!; Machakos District: Mua Hills, 31 Dec. 1971, *Kibue* 172!; Masai District: Ngong Hills, 6 Mar. 1965, *Kokwaro* 1!
TANZANIA. Arusha District: Ngurdoto National Park, 4 Nov. 1965, *Greenway* 12275!; Ufipa District: near Molo village, Malonje plateau, 1 Jan. 1962, *Richards* 15840!; Iringa District: Mufindi, 19 Jan. 1969, *Paget-Wilkes* 320!
DISTR. **U** 1–4; **K** 1–7; **T** 1–7; widespread in tropical Africa, tropical Asia and Australia
HAB. Miombo woodland, bushland, riverine forest or thicket, upland forest especially in rocky places; 0–2250 m
USES. Plant used as hedge; fruit edible; root decoction drunk as soup or as pain killer; minor medicinal use

SYN. *Antura edulis* Forssk., Fl. Aegypt.-Arab.: 63 (1775)
   Carissa edulis (Forssk.) Vahl in Symb. Bot. 1: 22 (1790); Stapf in F.T.A. 4, 1: 89 (1902); T.T.C.L.: 48 (1949); K.T.S.: 45 (1961); F.P.U.: 117 (1962); Huber in F.W.T.A. ed. 2, 2: 54 (1963); Codd in F.S.A. 26: 251 (1963); Kupicha in F.Z. 7, 2: 399 (1985); Wild Flow. E. Afr.: 142, t. 69 (1987); K.T.S.L.: 478, ill., map (1994); U.K.W.F. ed. 2: 171 (1994). Type: Yemen, Mt Hadiens, *Forsskål* s.n. (C, holo., BM, S, iso.)
   C. tomentosa A. Rich., Tent. Fl. Abyss. 2: 30 (1851). Type: Ethiopia, Tchelikote Province, "Agam", *Petit* s.n. (P, lecto., chosen by Leeuwenberg & van Dilst)
   C. richardiana Jaub. & Spach., Ill. Pl. Or. 5: t. 496 (1857). Type: Ethiopia, Tigray, near Adwa, *Schimper* 156 (P, lecto., BM, BP, BR, FT, G-DC, GH, K, L, LG, M, MEL, NY, P, S, UPS, W, WAG, Z, iso., chosen by Leeuwenberg & van Dilst)
   C. candolleana Jaub. & Spach., Ill. Pl. Or. 5: t. 497 (1857). Type: Ethiopia, Tigray, near Adwa, *Schimper* 209 (P, lecto., BM, BP, BR, FT, G-DC, HBG, K, M, MEL, W, iso., chosen by Leeuwenberg & van Dilst)
   C. cornifolia Jaub. & Spach., Ill. Pl. Or. 5: t. 498 (1857). Type: Ethiopia, Gondar, Senen, Sanfetch Mt, *Schimper* 1068 (P, lecto., BR, G, K, W, WAG, iso., chosen by Leeuwenberg & van Dilst)
   C. edulis (Forssk.) Vahl var. *major* Stapf in F.T.A. 4, 1: 89 (1902); T.T.C.L.: 48 (1949). Lectotype: Malawi, Manganja Hills, *Meller* s.n. (K, lecto., chosen by Kupicha)
   C. edulis (Forssk.) Vahl var. *tomentosa* (A. Rich.) Stapf in F.T.A. 4, 1: 89 (1902); T.T.C.L.: 48 (1949)

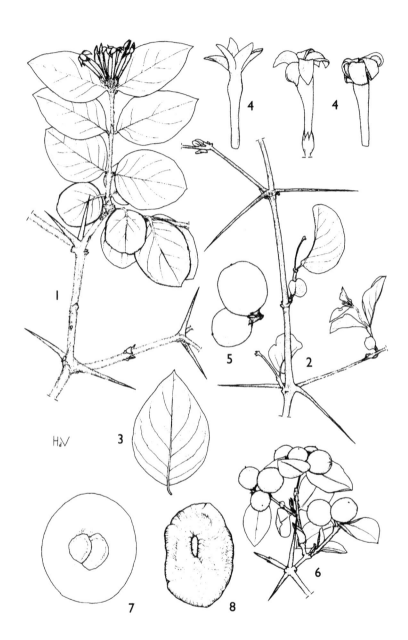

Fig. 4. *CARISSA SPINARUM* — **1**, flowering branch, × ²/₃; **2**, branch with spines, × ²/₃; **3**, leaf, × ²/₃; **4**, flower, × 2; **5**, fruit, × 1; **6**, fruiting branchlet, × 1; **7**, section of fruit, × 1.3; **8**, seed, × 5. 1 from *de Wilde* 5531; 2–3 from *Schmitz* 6670; 4 from *Leeuwenberg* 10787; 5 & 7–8 from *Leeuwenberg* 10100; 6 from *Westphal* 416. From Wageningen Univ. Papers 01.1, t. 8 (2001), drawn by H. de Vries, reproduced by permission.

3. **C. tetramera** (*Sacleux*) *Stapf* in F.T.A. 4, 1: 91 (1902); U.O.P.Z.: 173 (1949); T.T.C.L.: 48 (1949); K.T.S.: 45 (1961); Codd in F.S.A. 26: 252 (1963); Kupicha in F.Z. 7, 2: 400 (1985); K.T.S.L.: 478, map (1994); Leeuwenberg & van Dilst in Wageningen Univ. Pap. 01.1: 70, t. 11, map 8 (2001). Type: Tanzania, Zanzibar, Nov. 1847, *Boivin* s.n. (P, lecto., chosen by Leeuwenberg & van Dilst)

Shrub or tree 1–5 m high; bark brownish grey; spines furcate, sometimes bifurcate or simple, spine axis 0.8–3 cm long, branches 1–6 cm long. Leaves petiolate; petiole 1–2 mm long, usually 1–3 stipules or bracts present in each leaf axil; blade ovate, elliptic or suborbicular, 1–8 cm long, 0.4–6 cm wide, acute or rounded and mucronate at the apex, cuneate or rounded at the base, glabrous or pubescent, with 12–18 straight pairs of secondary veins. Inflorescence terminal, few-flowered. Flowers white inside, pink to red outside; sepals 1–3.4 mm long, ovate to lanceolate, acute to acuminate, cilate; corolla tube 5–19 mm long, lobes overlapping to the left, obovate or elliptic, 2–10 mm long, 2–6 mm wide, rounded, not ciliate; stamens inserted at ± the middle of the tube. Fruit red to black, subglobose or ellipsoid, 9–20 mm long, 7–10 mm in diameter, 4–8 seeded.

KENYA. Lamu District: Boni Forest Reserve, 11 km W of Mangai, 30 Nov. 1988, *Luke & Robertson* 1519!; Kilifi District: Mida, Aug. 1934, *Elliot* 752!; Kwale District: between Samburu and Mackinnon road, 5 Sept. 1953, *Drummond & Hemsley* 4179!
TANZANIA. Lushoto District: Daluni, Mahewa, 26 Oct. 1935, *Greenway* 4126!; Handeni District: near Sindeni, 16 Sept. 1971, *Bally* 14286!; Uzaramo District: Pugu Forest Reserve, 9 Mar. 1964, *Semsei* 3697!; Zanzibar: Masazini, Sep. 1959, *Faulkner* 2360
DISTR. **K** 7; **T** 3, 6, **Z**; Mozambique, Zimbabwe, Swaziland, South Africa
HAB. Dry coastal forest, thicket and bushland; 0–450 m
USES. Fruit edible; stems used for making bows

SYN. *Arduina tetramera* Sacleux in Journ. de Bot. sér. 1, 7: 312 (1893)

2. ACOKANTHERA

G. Don. in Gen. Syst. 4: 485 (1838); Kupicha in K.B. 37 (1): 45 (1982)

Trees or shrubs, all parts (except the ripe fruit) containing toxic cardiac glycosides. Leaves opposite and decussate, exstipulate. Flowers white tinged with red, borne in dense axillary cymes, sweetly scented. Sepals imbricate, free to base, ovate to lanceolate. Corolla hypocrateriform; tube cylindrical; lobes contorted, overlapping to the left, shorter than the tube; ovary superior, glabrous, bicarpellate, syncarpous with axile placentation; ovule 1 per locule; style filiform; stigma minutely bifid. Fruit a berry, containing one or two seeds.

A genus of 5 species confined to Africa; one species, *A. schimperi*, extending to Yemen.

1. Secondary veins of leaf clear and raised above; corolla tube
   12–19 mm long ·   3. *A. oppositifolia*
   Secondary veins faint, not raised above; corolla tube 8–11
   mm long · 2
2. Leaves more than 8 cm long; secondary veins more than 5;
   sepals 2.5–3.5 mm long ·   1. *A. laevigata*
   Leaves mostly less than 8 cm long; secondary veins fewer
   than 5; sepals 1–2 mm long · · · · · · · · · · · · · · · · · ·   2. *A. schimperi*

1. **A. laevigata** *Kupicha* in K.B. 37, 1: 58 (1982) & in F.Z. 7, 2: 406 (1985). Type: Tanzania, Kilwa District, Libungani, *Page Jones* H9/36 (K, holo., EA!, iso.)

Tree 5–12 m high with white sticky latex; bark pale brown, fissured; slash pale yellow, fibrous, wood white; branches glabrous; branchlets glabrous or puberulous. Leaves petiolate; petiole 5–6 mm long, glabrous or puberulous; blade elliptic, 4–13 cm long, 3.5–7.3 cm wide, cuneate or rounded at the base, obtuse to rounded or acute at the apex with a hard mucro, rarely emarginate, glabrous to puberulous especially on midrib beneath. Inflorescence of contracted axillary cymes; sepals ovate, 2.5–3.5 mm long, apiculate, ciliate otherwise glabrous. Corolla tube pink, 9.5–11 mm long, sparsely pubescent outside, pubescent in the upper half inside; lobes white, ovate-cuspidate, 3–4 mm long, glabrous, densely hairy at the base and mouth of the tube, ciliate; stamens just visible at anthesis; anthers 1 mm long, apiculate at the apex, apiculus 0.6–0.7 mm long; pistil 9–10 mm long. Fruit ovoid, 2.5–3 cm long, up to 2.5 cm in diameter.

TANZANIA. Bagamoyo District: Bagamoyo, 17 Sept. 1935, *Raymond* DSM 80!; Morogoro District: Liwale, June 1950, *Ionides* in *Bally* 7925!; Iringa District: Lupeme Tea Estate, 30 Aug. 1971, *Perdue & Kibuwa* 11307!
DISTR. **T** 6–7; Malawi
HAB. Wooded grassland, clumped tree grassland; 1–2000 m
USES. All parts of the plants are poisonous except the ripe fruit; used in arrow poison

2. **A. schimperi** (*A.DC.*) *Schweinf.* in Bol. Soc. Afr. Italia 10 (11–12): 12 (1891); Verdcourt & Trump, Comm. Poison. Pl. E. Afr.: 123, t. 8/e–f (1969); Kupicha in K.B. 37, 1: 48 (1982); Wild Flow. E. Afr.: 142, t. 68 (1987); K.T.S.L.: 475 (1994). Type: Ethiopia, Ado, near Schahagenni, *Schimper* 254 (P, holo., BM, BR, CGE, FT, FI-W, K, iso.)

Shrub or tree 1–10 m high, branching from the base; bark rough grey-brown; slash cream with fine orange streaks; young branches glabrous or pubescent, conspicuously angled and ribbed. Leaves petiolate, petiole 3–6 mm long; blade ovate, obovate, elliptic to suborbicular, 1–8(–10) cm long, 1.1–6.5 cm wide, rounded or acute at the apex, with a hard mucro, sometimes emarginate, cuneate or rounded at the base, coriaceous, glabrous; secondary veins obscure, especially beneath. Inflorescence contracted, of many-flowered axillary cymes; sepals ovate, 1–2 mm long, 0.5–1 mm wide, pubescent or glabrous, ciliate. Corolla tube pink or reddish, 8–11 mm long, glabrous or pubescent on external surface, inner surface pilose in the upper half and wrinkled below; lobes pink outside, white inside, ovate, 2.5–3.5 mm long, 1.5–2.5 mm wide, acuminate at the apex, pubescent at the base inside, glabrous or pubescent outside, ciliate; stamens slightly exserted, inserted at 7–10 mm from the base; anthers ovate, 1–1.5 mm long, acumen 0.2 mm long; pistil 7.5–10 mm long. Fruit purple when ripe, ellipsoid, 10–25 mm long; seeds up to 13 mm long. Fig. 5/5–6 (p. 16).

UGANDA. Karamoja District: Lokitonyala, Mar. 1960, *Wilson* 800!; Mbale District: Siti R., Jan. 1948, *Eggeling* 5725! & Elgon, Oct. 1939, *Dale* 66
KENYA. Masai District: Ngong Hills, Mar. 1950, *Bally* 7771!; Kiambu District: Kikuyu Escarpment Forest, 16 Sept. 1979, *Kuchar* 12251!; Baringo District: Tenges, 26 Jul. 1970, *Gibson* 4!
TANZANIA. Musoma District: between Kuka and Nyamalumba Ranges, 15 Oct. 1961, *Greenway & Turner* 10269!; Masai District: Loliondo Forest Reserve, 8 Nov. 1968, *Carmichael* 1602!; Mbulu District: Aitcho Mts, 31 Aug. 1932, *Burtt* 4264!
DISTR. **U** 1, 3; **K** 1–7; **T** 1, 2; Congo (Kinshasa) and Rwanda, Eritrea, Ethiopia, Djibouti, Somalia; Yemen
HAB. Dry forest and forest margins, wooded grassland, rocky bushland; 250–2200 m
USES. All parts of the plants are poisonous except the ripe fruit, which is edible; bark and wood used in one of most important arrow poisons

FIG. 5. *ACOKANTHERA OPPOSITIFOLIA* — **1**, flowering shoot, × $\frac{1}{2}$; **2**, fruiting shoot, × $\frac{1}{2}$; **3**, flower, × 2; **4**, seed, × 1. *ACOKANTHERA SCHIMPERI* — **5**, fruiting shoot, × $\frac{1}{2}$; **6**, flower, × 3. From Verdcourt & Trump, Poisonous plants of East Africa, drawn by M.E. Church.

SYN. *Carissa schimperi* A. DC. in DC., Prodr. 8: 675 (1844)
    Acokanthera ouabaio Lewin in E.J. 17, Beibl. 41: 44 (1893), nom. nud. based on Somali
    specimen
    A. abyssinica K. Schum. in P.O.A. C: 315 (1895), nom. superfl. illeg. pro *Carissa schimperi*
    A. friesiorum Markgr. in N.B.G.B. 8: 466 (1923); T.T.C.L.: 47 (1949); I.T.U.: 24 (1952);
    K.T.S.: 48 (1961). Lectotype: Kenya, N Nyeri District, Coles Mill, *R & T Fries* 1127 (K,
    lecto., chosen by Kupicha)
    A. wabajo Markgr. in N.B.G.B. 8: 465 (1923). Lectotype: Somalia, Heid, Ahl Mts, *Hildebrandt*
    1431 (K, lecto., BM, iso., chosen by Kupicha)
    A. deflersii Lewin var. *africana* Markgr. in N.B.G.B. 8: 464 (1923). Lectotype: Ethiopia,
    Ghinda near R. Telekwit, *Schweinfurth & Riva* 2183 (K, lecto., P, iso., chosen by Kupicha)
    A. deflersii Lewin var. *africana* Markgr. forma *scabriuscula* Markgr. in N.B.G.B. 8: 464 (1923).
    Type: Ethiopia, between Ghinda and Filagabai, *Schweinfurth* 528 (not found)
    Carissa friesiorum (Markgr.) Cufod. in B.J.B.B. 30 suppl.: 684 (1960)

3. **A. oppositifolia** (*Lam.*) *Codd* in Bothalia 7: 448 (1961) & in F.S.A. 26: 247, t.
36/1 (1963); Kupicha in K.B. 37, 1: 53, fig. 1 (1982) & in F.Z. 7, 2: 405, t. 91 (1985);
K.T.S.L.: 475 (1994). Type: Africa, *Sonnerat* (P-LA, holo.)

Shrub or small tree 1–7 m high; bark brown, deeply fissured; slash cream turning
olive green. Leaves petiolate; petiole 2–7 mm long; blade elliptic or obovate, 4–13.5
cm long, 1.5–8.4 cm wide, acute, obtuse or rounded at the apex, with a hard spiny
mucro, cuneate or rounded at the base, coriaceous, glabrous, sometimes puberulous
on young leaves especially beneath on lower nerves; secondary veins prominent.
Inflorescence of dense cymes; bracts ovate, 1.3 mm long, acute at the apex, ciliate;
sepals ovate, 2.4–3.5 mm long, 0.9–1.6 mm wide, acuminate to acute at the apex,
dorsally pubescent or glabrous, ciliate. Corolla tube pink or red, 12–19 mm long,
pubescent or hispid on external surface, rarely glabrous, pilose in the upper half and
wrinkled below inside; lobes broadly elliptic, 3.2–4 mm long, 2.5–4 mm wide, acute
at the apex, pubescent at the base and mouth of tube, ciliate; stamens slightly
exserted, inserted at 12.5–15 mm from the base; anthers ovate up to 1.3 mm long;
pistil 13–18 mm long; style 11–16 mm long. Fruit purple, ellipsoid, 12–35 mm long,
± 25 mm in diameter; seeds 6.6–10 mm long. Fig. 5/1–4 (p. 16).

KENYA. Kiambu District: Karura Forest, 9 Dec. 1969, *Perdue & Kibuwa* 10181!; Machakos
    District: 20 miles NE of Machakos, 20 Oct. 1967, *Mwangangi* 272!; Teita District: on the road
    to Ngangao forest, 22 Sept. 1989, *Omino* 72!
TANZANIA. Arusha District: Tengeru, Aug. 1986, *van der Laan* 1200!; Lushoto District: Shume
    forest, Nov. 1957, *Semsei* 2721!; Mbeya District: 25 miles S of Mbeya, Sept. 1959, *Procter* 1357!
DISTR. **K** 4, 7; **T** 2, 3, 7; Congo (Kinshasa), Zambia, Malawi, Mozambique, Zimbabwe, Swaziland
    and South Africa
HAB. Dry forest and forest margins, riverine forests and thickets, occasionally in woodland;
    1200–2200 m
USES. Roots and bark used in arrow poison; fruits said to be edible or poisonous

SYN. *Cestrum oppositifolium* Lam., Tab. Encycl. 1, 1 (2): 5, t. 112, t. 2 (1792) & 2, 3 (1): 5 (1794)
    Acokanthera longiflora Stapf in K.B. 1922: 28 (1922); T.T.C.L.: 47 (1949); K.T.S.: 43 (1961);
    Verdcourt & Trump, Comm. Poison. Pl. E. Afr.: 123, t. 8/a–d (1969). Lectotype: Tanzania,
    Lushoto District, Kwa Msusha, *Holst* 8968 (K, lecto., BM, isolecto., chosen by Kupicha)
    Carissa oppositifolia (Lam.) Pichon in B.J.B.B. 22: 109 (1952)
    C. longiflora (Stapf) Lawrence in Baileya 7: 90 (1959)
    Garcinia sciura Spirlet in B.J.B.B. 29: 327 (1959). Type: Congo (Kinshasa), Shaba [Katanga],
    5 km S of Lubumbashi [Elisabethville], *Schmitz* 1368 (BR, holo.)
    Acokanthera venenata sensu T.T.C.L.: 47 (1949), *non* G. Don

NOTE. This species is also cultivated in the Nairobi Arboretum and at Amani.

3. LANDOLPHIA

P. Beauv., Fl. Owar.1, 6: 54 (1804), nom. conserv.; Pichon in Mém. I.F.A.N. 35: 40 (1953); Persoon et al. in Wageningen Agric. Univ. Pap. 92, 2: 6–209 (1992)

Aphanostylis Pierre in Bull. Soc. Linn. Paris sér. 2: 89 (1898)

Lianas, rarely sarmentose shrubs with long curled tendrils, branched at the very tip; latex present in all parts, often abundant; stipules absent. Leaves opposite; petiole often alternating with one or several pairs of bracts. Inflorescence terminal and/or axillary, sometimes tendril-like, a lax to dense cyme. Flowers 5-merous, fragrant. Sepals imbricate, free or shortly united at the base. Corolla variously coloured; tube cylindrical, slightly enlarged at the level of the anthers, thickened above the anthers; lobes broadly elliptic to narrowly linear, contorted, overlapping to the left; stamens included, free from each other and from pistil head; filaments very short; ovary glabrous or hairy; ovules numerous; style terete, stigma bifid, subtended by a clavuncula positioned at the level of the base of the anthers. Fruit a many-seeded indehiscent berry with or without a sclerified pericarp; pulp sweet and acid. Seeds many, endospermous.

A genus of 50 species in Africa and 10–13 species in Madagascar.

Note: *Landolphia* spp. produce abundant latex, used for rubber and bird lime. The fruits of most species are edible.

Though F.W.T.A. ed. 2, 2: 57 (1963) states that *L. congolensis* (Stapf) Pichon occurs in Uganda, this is not true. T.T.C.L.: 51 (1949) lists *L. heudelotii* A. DC. as an Tanzanian species; this is a W African species, and has been cultivated at Amani (without date, *Zimmermann* s.n.; Jan. 1943, *Greenway* 6663). Similarly, *L. klainei* Pierre was grown at Amani for its rubber (T.T.C.L.: 52 (1949); *L. mannii* Dyer was also cultivated at Amani (June 1929, *Greenway* 1583; Feb. 1943, *Greenway* 6668).

1. Inflorescence predominantly axillary; petiole and blade
   glabrous · 2
   Inflorescence terminal, sometimes also axillary (*L. buchananii*); petiole and blade puberulous to tomentose or sometimes glabrous (in *L. buchananii, owariensis* and *parvifolia*) · 3
2. Leaves less than 12 cm long, distinctly acuminate; inflorescence 2–11-flowered; corolla tube < 5 mm long; coastal Tanzania and Kenya · · · · · · · · · · · · · · · ·   7. *L. watsoniana*
   Leaves usually > 12 cm long, obtusely acuminate to obtuse; inflorescence 15–45-flowered; corolla tube > 8 mm long; Uganda ·   4. *L. landolphioides*
3. Corolla tube 13–24 mm long · · · · · · · · · · · · · · · · · · ·   2. *L. eminiana*
   Corolla tube 2.3–12 mm long · 4
4. Inflorescence 2–20-flowered; sepals glabrous outside; corolla densely pubescent at mouth and base of lobes · ·   1. *L. buchananii*
   Inflorescence usually more than 20-flowered; sepals pubescent outside; corolla glabrous at mouth and base of lobes · 5
5. Leaves mostly > 10 cm long; secondary veins prominent, curved towards the margin, more than 15 mm apart at median part of leaf; corolla tube > 6 mm long · · · · · ·   5. *L. owariensis*
   Leaves mostly < 10 cm long; secondary veins less prominent, straight, less than 10 mm apart at middle part of leaf; corolla tube < 5.5 mm long · 6

6. Sepals densely ciliate, pubescent outside on the midrib, otherwise glabrous and glistening reddish brown; ovary pubescent ··································   6. *L. parvifolia*

    Sepals less densely ciliate, pubescent over the whole surface outside; ovary glabrous or occasionally sparsely pilose near apex ····························   3. *L. kirkii*

1. **L. buchananii** (*Hallier f.*) *Stapf* in F.T.A. 4, 1: 35 (1902); T.T.C.L.: 51 (1949); Pichon in Mém. I.F.A.N. 35: 48 (1953); Kupicha in F.Z. 7, 2: 409 (1985); Persoon in Wageningen Agric. Univ. Pap. 92, 2: 27, t. 2 (1992); K.T.S.L.: 481, ill., map (1994). Type: Malawi, Shire Highlands, *Buchanan* 220 pro parte (B†, holo., BM, lecto., K!, isolecto., chosen by Persoon)

Liana 3–40 m high, rarely sarmentose shrub 0.9–7 m high with copious white latex; stems deeply fluted, to 22 cm in diameter, dark brown; branchlets glabrous or pilose. Leaves petiolate, blade elliptic to ovate or obovate, 1.9–14.5 cm long, 0.8–6 cm wide, obtuse to bluntly acuminate at the apex with acumen up to 1 cm long, rounded to cuneate at the base, sometimes slightly ciliate, glabrous above, lower surface completely glabrous or pubescent at the base of midrib and lower secondary veins; petiole 1.5–8 mm long, glabrous or pilose, with 4–8 triangular glands on each node, often with empty bracts in between nodes. Inflorescence terminal, sometimes also axillary, 1.6–5.5 cm long, 2–20-flowered; peduncle 5–33 mm long, sometimes tendril-like, elongate and curved; bracts ovate to triangular, 1–2.5 cm long; pedicels 1.2–6.5 mm long. Flowers fragrant; sepals ovate, 1.2–3.1 mm long, rounded at the apex; corolla white, creamy or yellow; tube often greenish, occasionally tinged pink or reddish, 6–12 mm long; lobes narrowly ovate to elliptic, 5.3–15.5 mm long, 1.9–4 mm wide, glabrous, rounded; stamens inserted 1.8–3.6 mm from the base of the tube, included for 1.3–5.9 mm; ovary ovoid, glabrous or with a few hairs. Fruits green with white or light brown verrucose spots, 1–50 mm in diameter, globose or pyriform, 2.5–6 cm long, 2–20-seeded; pericarp 5–6.5 mm thick; pulp fleshy, white, turning yellow when exposed; seed up to 17 mm long, irregularly ovoid, laterally compressed, minutely pitted.

UGANDA. Ankole District: Ruizi R., 15 May 1950, *Jarrett* 315!; Bunyoro District: Bugoma Forest, Feb. 1933, *Brasnett* 1315!; Masaka District: Maligambo Forest 6.5 km SSW of Katera, 2 Oct. 1953, *Drummond & Hemsley* 4574!
KENYA. Fort Hall District: below Thika Falls near junction with Chania R., 2 Nov. 1968, *Faden* 68/811!; South Kavirondo District: Kisii, between Irigonga and Ogumo, 27 Dec. 1974, *Vuyk* 426!; Masai District: Namanga R. camp, 20 May 1945, *Bally* 4516!
TANZANIA. Bukoba District: Minziro Forest, 18 Oct. 1953, *Willan* 37!; Lushoto District: Amani, near Forest House, 24 Sept. 1967, *Harris* 1039!; Mpanda District: Kabwe R., S of Pasagulu Mt, 8 Aug. 1959, *Harley* 9195!
DISTR. **U** 2–4, **K** 1, 3–7, **T** 1–5, 7; Nigeria, Congo (Kinshasa), Burundi, Sudan, Ethiopia, Somalia, Angola, Zambia, Malawi, Mozambique and Zimbabwe
HAB. Riverine or swamp forest; 450–2400 m
USES. Stem used as rope; latex for rubber; fruit edible

SYN. *Clitandra buchananii* Hallier f. in Jahrb. Hamb. Wiss. Anst. 17, Beih. 3: 119 (1900)
    C. kilimanjarica Warb. in Tropenfl. 4: 614 (1900). Lectotype: Tanzania, Kilimanjaro, Moshi, *Merker* 224 (K!, lecto., chosen by Pichon)
    Landolphia kilimanjarica (Warb.) Stapf in F.T.A. 4, 1: 34 (1902); T.T.C.L.: 51 (1949); U.K.W.F. ed. 2: 171 (1994), as *kilimandjarica*
    L. cameronis Stapf in F.T.A. 4, 1: 35 (1902). Lectotype: Malawi, Namasi, *Cameron* 1 (K, lecto., chosen by Persoon)
    L. ugandensis Stapf in F.T.A. 4, 1: 589 (1904); T.T.C.L.: 52 (1949). Type: Uganda, Masaka District, Dumu Forest, Buddu, *Dawe* 23 (K!, holo.)
    L. rogersii Stapf in K.B. 1912: 360 (1912). Type: Congo (Kinshasa), Lubumbashi, *Rogers* 10078 (K, holo.)
    Clitandra semlikiensis Robyns & Boutique in B.J.B.B. 18: 259 (1947). Type: Congo (Kinshasa), Djalele R., W of Katuka, *De Wilde* 2 (BR, holo.)

Jasminochyla ugandensis (Stapf) Pichon in Bull. Mus. Hist. Nat., sér. 2, 20: 551 (1949)

NOTE. Cultivated at Amani.

2. **L. eminiana** *Hallier f.* in Jahrb. Hamb. Wiss. Anst. 17, Beih. 3: 88 (1900); T.T.C.L.: 51 (1949); Pichon, Mém. I.F.A.N. 35: 137 (1953); Kupicha in F.Z. 7 (2): 413 (1985); Persoon in Wageningen Agric. Univ. Pap. 92, 2: 60, t. 9 (1992). Lectotype: Tanzania, Bukoba District, Itolio, *Stuhlmann* 930 (K!, lecto., P, isolecto., chosen by Pichon)

Liana or sarmentose shrub 3–20 m high; branches brown, with light lenticels; branchlets greyish to reddish brown, pubescent or lanate with brown to orange indumentum. Leaves petiolate, blade elliptic, oblong to obovate, 2.1–9(–15) cm long, 0.8–4.8 cm wide, obtuse to acuminate at the apex with acumen up to 1.5 mm long, cuneate to rounded at the base, densely pilose or lanate on midrib; petiole 3–10 mm long, sparsely to densely pubescent. Inflorescence terminal, sometimes tendril-like, 3.5–6 cm long, up to 15 cm long when tendril-like, 1–40-flowered; peduncle up to 2.8 cm long, tendril hooks up to 4.3 cm long, pubescent or glabrous; pedicels up to 6 mm long, pubescent or tomentose. Flowers fragrant; sepals ovate or obovate, 2.3–7 mm long, pubescent on both sides, with a ciliate margin; corolla white or greyish white, less often orange or pale yellow; tube 13–24 mm long; lobes narrowly ovate or elliptic, 8.4–15 mm long, 3–7.8 mm wide, obtuse at the apex; stamens inserted at 8–12 mm from the base of the corolla tube, included for 3.5–7.4 mm; pistil 8.4–11.5 mm long; ovary villose, cylindrical, base glabrous, ovules 60–120. Fruits orange, with light or dark lenticels, globose to pyriform, 3–10 cm in diameter, rounded or truncate at the apex, 5–15-seeded; pericarp 1.5–5 mm thick; scleroid cell layer 0.3–0.5 mm thick; seeds up to 7 mm long.

KENYA. Kilifi District: Kaya Jibana, *Lap* 307!
TANZANIA. Bukoba District: 88 km S of Bukoba, 1958, *Procter* 1032!; Kigoma District: Lugufu, 16 Sept. 1942, *Perry* H61/42!; Iringa District: Mafinga, 19 Jul. 1980, *Leedal* 5972!
DISTR. **K** 7? (see note); **T** 1, 4, 7; Congo (Kinshasa), Rwanda, Burundi, Zambia and Malawi
HAB. Riverine and lakeside forest, woodland and thicket; (0–)1100–1350 m
USES. Latex used for rubber; fruit edible

NOTE. The Kenyan specimen is poor and without adequate notes; it is possibly misplaced here.

3. **L. kirkii** *Dyer* in Rep. Roy. Gard. Kew 1881: 39 (1881); U.O.P.Z.: 323, ill. (1949); T.T.C.L.: 52 (1949); Pichon in Mém. I.F.A.N. 35: 88 (1953); Codd in F.S.A. 26: 259 (1963); Kupicha in F.Z. 7 (2): 411 (1985); Persoon in Wageningen Agric. Univ. Pap. 92 (2): 102, t. 18, (1992); K.T.S.L.: 481, map (1994). Type: Tanzania, Uzaramo District, Dar es Salaam, *Kirk* s.n., Oct. 1868 (K!, lecto., chosen by Persoon)

Liana 1–30 m long or erect or straggling shrub 0.3–4(–8) m high, with white latex in all parts; trunk 2–15(–30) cm in diameter; bark dark brown or greyish purple, smooth or rough; branchlets with whitish, green or brown indumentum. Leaves petiolate, blade elliptic to ovate, 0.9–9(–10) cm long, 0.3–4.4 cm wide, retuse to bluntly acuminate at the apex, cuneate to rounded at the base, pilose to tomentose or glabrous, margin ciliate or not; petiole 2–7 mm long, pilose or tomentose, sometimes glabrous, often with glands in the axils, alternating with one or several pairs of bracts. Inflorescence terminal or at the end of tendrils hooks, 0.8–5.5 cm long, up to 14 cm long in tendrils, 3–70-flowered, pubescent in all parts; peduncle 2–45 mm long; bracts triangular or ovate, 0.5–2.5 cm long; pedicel 0.5–3 mm long. Flowers fragrant or not; sepals ovate or obovate, 0.9–3.5 mm long, pubescent outside; corolla white or cream, buds yellow; tube pilose to glabrous, 2.3–4.2 mm long; lobes, ovate or narrowly ovate, 2.1–6 mm long, 0.7–2.3 mm wide, rounded to mucronate at the apex; stamens inserted 1.2–2.5 mm from the base, included for 0.1–1.2(–1.8) mm;

pistil (1.5–)2–3.1 mm long; ovary ovoid to conical, glabrous, occasionally sparsely pilose in the upper part; ovules (30–)40–70(–100); style cylindrical, (0.15–)0.3–1.1 mm long, glabrous. Fruit dull green, spotted yellow, globose or pyriform, 5–15 cm long and 5 cm in diameter, 10–30 seeded; pericarp up to 8 mm thick; pulp orange, red or yellow.

UGANDA. Toro District: Ruwenzori Mts, Stevenson Rock, *Scott Elliot* 8375
KENYA. Lamu District: Boni Forest Reserve, 8 km E of Mangai, 25 Nov. 1988, *Luke & Robertson* 1508!; Kwale District: Mrima Hill, 7 Dec. 1975, *Kokwaro* 3960!; Kilifi District: Mwarakaya, 4.9 km S of Kilifi–Kaloleni road, 6 May 1985, *Faden & Beentje* 85/15!
TANZANIA. Ufipa District: Muva Forest Reserve, Sumbawanga, 7 Nov. 1963, *Carmichael* 994!; Morogoro District: Mgolole, near Morogoro, Nov. 1951, *Eggeling* 6356; Lindi District: Lutamba, 8 Sept. 1934, *Schlieben* 5271!
DISTR. **U** 2; **K** 7; **T** 2–4, 6–8; **Z**; **P**; Central African Republic, Congo (Kinshasa), Zambia, Malawi, Mozambique, Zimbabwe and South Africa
HAB. Forest and woodland, bushland or thicket, also in wooded grassland on or near rocky outcrops; 0–1600 m
USES. Stem used as string; latex for rubber and as minor medicine; leaves as vegetable; reports of fruit being poisonous or pulp being edible

SYN. *L. kirkii* Dyer var. *delagoensis* Dew. in Ann. Soc. Sci. Brux. 19 (2): 140 (1895). Type: Mozambique, Delagoa Bay, *Junod* 101 (G, holo.)
    L. delagoensis (Dew.) Pierre in Bull. Soc. Linn. Paris, ser. 2: 15 (1898)
    L. polyantha K. Schum. in E.J. 28: 452 (1900). Type: Tanzania, Uzaramo District, Pugu Mts, *Goetze* 5 (B†, holo., K!, lecto., BM, BR, E, L, P, iso., chosen by Persoon)
    L. dondeensis Busse in Tropenfl. 5: 405 (1901). Lectotype: Tanzania, Kilwa District, Barikiwa–Donde, *Busse* 584 (P, lecto., BM, G, HBG, K!, L, W, isolecto., chosen by Pichon)
    L. kirkii Dyer var. *dondeensis* (Busse) Stapf in F.T.A. 4, (1): 56 (1902); T.T.C.L.: 52 (1949)

4. **L. landolphioides** (*Hallier f.*) *A.Chev.* in Rev. Bot. Appliq. 28: 401 (1948); Pichon in Mém. I.F.A.N. 35: 53 (1953); Huber in F.W.T.A. ed. 2, 2: 56 (1963); Persoon in Wageningen Agric. Univ. Pap. 92, 2: 112, t. 20 (1992). Type: Cameroon, Buea, *Deistel* 556 (B†, holo., P, lecto., A, BM, HBG, K, MO, S, UPS, US, W, Z, isolecto., chosen by Pichon)

Liana 5–30 m high; trunk up to 30 cm in diameter; branchlets tomentose or glabrous. Leaves petiolate, blade obovate or elliptic, 6–26 cm long, 2.8–10.3 cm wide, obtuse to obtusely acuminate at the apex, acumen up to 1 cm long, rounded to acute at the base, glabrous; petiole 4–17 mm long, glabrous, sometimes with glands in the axil. Inflorescence predominantly axillary, sometimes also terminal, 2.5–6.5 mm long, 3–45-flowered; peduncle 3–15 mm long, glabrous to tomentose; pedicels 2–8 mm long, glabrous to tomentose. Flowers fragrant; sepals ovate, 1.4–3 mm long, rounded to obtuse at the apex; corolla white, creamy or yellow, often differently coloured on tube; tube 8–14 mm long; lobes 6.5–17.4 mm long, 2.5–4.3 mm wide, rounded at the apex; ovary glabrous to tomentose, ovules 50–180. Fruits yellow or glaucous, often with scattered lenticels, pyriform, subglobose or ovoid, 3.5–8 cm long, 2.5–7 cm in diameter, pulp fleshy, yellowish, 5–30 seeded; pericarp 0.4–7.5 mm thick; seeds irregularly ovoid, up to 21 mm long, laterally compressed, minutely pitted.

UGANDA. Karamoja District: Loro Forest, near Kipandi R., *Dawe* 504; Masaka District: Sese Is., 4 Dec. 1942, *Thomas* 4100!; Mengo District: Kipayo Estate, *Dummer* 2431
DISTR. **U** 1, 2, 4; Guinea, Sierra Leone, Nigeria, Cameroon, Gabon, São Tomé, Central African Republic, Congo (Kinshasa), Sudan and Angola
HAB. Moist forest and riverine forest; 1150–1500 m
USES. Latex used for rubber; fruit edible

SYN. *Clitandra landolphioides* Hallier f. in Jahrb. Hamb. Wiss. Anst. 17, Beih. 3: 119 (1900)
    Carpodinus landolphioides (Hallier f.) Stapf in F.T.A. 4, 1: 80 (1902)
    Landolphia dawei Stapf in F.T.A. 4, 1 590 (1904). Lectotype: Uganda, Masaka District, Dumu Forest, Buddu, *Dawe* 47 pro parte (K!, lecto., chosen by Persoon)

5. **L. owariensis** P. *Beauv.* in Fl. Owar. 1, 6: 55 (1804); T.T.C.L.: 52 (1949); Pichon in Mém. I.F.A.N. 35: 53 (1953) pro parte; Huber in F.W.T.A. ed. 2, 2: 55, t. 213 (1963); Kupicha in F.Z. 7, 2: 412 (1985); van Dilst in Wageningen Agric. Univ. Pap. 92, 2: 153, t. 32 & 33 (1992). Type: Nigeria, Bendel State, Warri, *Palisot de Beauvois* s.n. (G, holo.)

Liana up to 100 m long, less often a shrub up to 3 m high; trunk up to 40 cm in diameter; bark grey-brown, rough, smooth or fissured, with or without protuding lenticels; branchlets brown pubescent, sometimes glabrous. Leaves petiolate, blade ovate or obovate to narrowly oblong, 2.1–16(–28) cm long, 1.1–6 cm wide, acuminate to obtuse at the apex, cuneate to rounded at the base, glabrous to puberulous on both sides, always pubescent to puberulous on midrib and secondary veins; petiole 1–12 mm long, pubescent, sometimes glabrous. Inflorescence terminal or on tendrils, in many-flowered lax to congested cymes; peduncle 0–15 cm long, pubescent, sometimes glabrous; pedicels 0.5–2.8 mm long. Flowers fragrant; sepals appearing red due to pubescence outside, ovate to oblong, 1.2–4 mm long, 1.5–3.2 mm wide, ciliate; corolla white or creamy, rarely yellow, turning pink to red after anthesis; tube urceolate to almost cylindrical, 6–9.9 mm long, variably puberulous outside; lobes oblong, 1.7–9.9 mm long, 1–3.2 mm wide, auriculate at the right side; stamens included for 0.3–2 mm, inserted 3–6.3 mm above the corolla base; ovary with a pubescent to lanate belt 0.4–1.1 mm at the retuse apex. Fruit yellow-orange or purple-blue, globose to broadly ovoid, 2.2–15 cm long, 1.8–15 cm in diameter, smooth or rough, sclerified, often lenticellate, pulp white or yellow sometimes orange tinged, many-seeded; fruit pericarp 0.2–15 mm thick; seeds brown to red brown, up to 18 mm long. Fig. 6 (p. 23).

UGANDA. Bunyoro District: Budongo Forest, *Synnott* 510; Masaka District: Bugala, Sese Is., 1 Feb. 1934, *Thomas* 1223! & Mugaye, Sese Is., 4 Dec. 1942, *Thomas* 4100!
TANZANIA. Buha District: Kakondo, *Tanner* 6010; Kigoma District: Kasakati, *Suzuki* 292; Rungwe District: Kyimbila, Beth Langern Mt, Mar. 1908, *Stolz* 7719!
DISTR. **U** 2, 4; **T** 4, 7; Senegal, Mali, Guinee Bissau, Guinea, Sierra Leone, Liberia, Ivory Coast, Ghana, Togo, Benin, Nigeria, Chad, Cameroon, Gabon, Congo (Brazzaville), Central African Republic, Congo (Kinshasa), Rwanda, Burundi, Sudan, Angola, Zambia and Malawi
HAB. Forest and forest margins, riverine forest, thickets in wooded grassland; 800–1550 m
USES. Latex used for inferior quality rubber; fruit edible

SYN. *Pacouria owariensis* (P. Beauv.) Hiern in Cat. Afr. Pl. Welw.1: 661 (1898)
   Landolphia gentilii De Wild., Obs. Apoc. Latex Rec. Gentil: 10 (1901). Type: Congo (Kinshasa), between Tumba and Leopold II Lakes, *Gentil* 1 (BR, holo.)
   L. owariensis P. Beauv. var. *tomentella* Stapf in F.T.A. 4, 1: 51 (1902). Type: Sudan, Djur land, Zeriba Ghattas, *Schweinfurth* 1377 (B†, holo., K, lecto., P, S, WU, iso., chosen by Persoon)
   L. stolzii Busse in E.J. 32: 168 (1902); T.T.C.L.: 52 (1949). Type: Tanzania, Rungwe District, Kondeland, *Stolz* 94 (K!, lecto., A, E, P, isolecto.)
   L. droogmansiana De Wild. in Lianes Caoutch. Et. Ind. Congo: 59 (1904). Type: Congo (Kinshasa), Kandakanda, *Gentil* 63 (BR, holo.)
   L. turbinata A.Chev. in Bull. Soc. Bot. Fr.53: 32 (1906). Type: Uganda, Masaka District, Dumu Forest, *Dawe* 47 pro parte (K!, holo.); the other part is *L. dawei* which is a synonym of *L. landolphioides*
   L. subturbinata Dawe in Rep. Bot. Miss. Uganda Protect.: 50 (1906). Type: as for *L. turbinata*
   L. kirkii Dyer var. *owariensis* (P. Beauv.) H. Durand in B.J.B.B. 2: 696 (1910)
   L. owariensis P. Beauv. var. *gentilii* (De Wild.) A.Chev. in Rev. Bot. Appliq. 28: 397 (1948)
   L. owariensis P. Beauv. var. *droogmansiana* (De Wild.) A.Chev. in Rev. Bot. Appliq. 28: 397 (1948)

NOTE. For a full synonymy see Persoon, 1992.
   The indumentum of the flowers is very variable even within a single inflorescence. The corolla lobes vary from being narrow to broad and from long to short, even within a single flower. The type specimen has calyces ranging from glabrous to densely hairy.

FIG. 6. *LANDOLPHIA OWARIENSIS* — **1**, flowering shoot × $^1/_2$; **2**, flower, × 1; **3**, section of flower, × 1.5; **4**, stamens × 4; **5**, fruit, × $^1/_3$. From F.W.T.A., drawn by S. Ross-Craig.

6. **L. parvifolia** *K. Schum.* in E.J. 15: 409 (1893); T.T.C.L.: 52 (1949); Kupicha in F.Z. 7, 2: 412 (1985); Persoon in Wageningen Agric. Univ. Pap. 92, 2: 162, t. 34 (1992). Type: Angola, Huila, Lopollo, *Weltwitsch* 5928 (B†, holo, BM, lecto., BR, C, COI, G, K!, P, isolecto., chosen by Persoon)

Liana up to 15 m or (scrambling) shrub or small tree, 1–10 m high, with or without curled tendrils up to 40 cm long; bark smooth, grey to brown; branchlets densely pilose to glabrous with redddish indumentum. Leaves petiolate, blade elliptic-oblong, 0.6–6.6 cm long, 0.4–2.7(–3.7) cm wide, shortly acuminate to rounded at the apex, subcordate to rounded at the base, glabrous to densely pilose; petiole 2–6 mm long, tomentose to glabrous. Inflorescence terminal, often with empty bracts, 1–4.5(–14) cm long, 2–40(–60)-flowered, pilose to glabrous; bracts resembling sepals, up to 3 mm long; peduncle 0.3–2.5 cm long, when tendril-like up to 6 cm long, with hooks up to 2.5 cm long; pedicels 0.2–1.2 mm long. Flowers fragrant; sepals reddish brown, ovate to obovate, 1.4–4 mm long, 0.9–2.9 mm wide, pubescent outside on midrib and margins; corolla yellowish white or pink, throat and buds often darker; tube 2.9–5.3 mm long; lobes ovate, elliptic or oblong, 2.2–7.9 mm long, 1–2.1 mm wide, rounded or obtuse at the apex, ciliate or not; stamens included for 0.2–0.9 mm, inserted 1.7–3 mm from the base; pistil 2.4–3.3 mm long; ovary ellipsoid to obovoid, pubescent; ovules 40–90; style 0.5–1.2 mm long. Fruits globose, whitish grey, yellow or greenish-violet, often with rusty to yellowish spots, up to 4.8 cm long, 8–30 seeded; pericarp 0.9–4 mm thick.

UGANDA. Kigezi District: Kinkizi Co., 7 km SW of Kirima, *Lye et al.* 4184!
TANZANIA. Dodoma District: 32 km S of Itigi on the Chunya road, 16 April 1964, *Greenway & Polhill* 11596!; Mbeya District: Ruaha National Park, summit of Magangwe Hill, 12 Mar. 1973, *Bjørnstad* 2625!; Songea District: near Mtandazi, ± 8 km W of Gumbiro, 26 Jan. 1956, *Milne-Redhead & Taylor* 8441!
DISTR. **U** 2; **T** 2, 4–8; Congo (Kinshasa), Burundi, Angola, Zambia, Malawi and Mozambique
HAB. Woodland, thicket and secondary bushland; 60–2000 m
USES. None recorded

SYN. *Pacouria parvifolia* (K. Schum.) Hiern in Cat. Afr. Pl. Welw.1: 663 (1898)
   Landolphia owariensis P. Beauv. var. *parvifolia* (K. Schum.) Stapf in F.T.A. 4, 1: 58 (1902)
   L. claessensii De Wild. in B.J.B.B. 5: 104 (1915). Type: Congo (Kinshasa), Sankuru, Katako-
   kombe, *Claessens* 380 (BR, holo.)

7. **L. watsoniana** *Romburgh* in Treub., Bot. Gart. s'Lands Plantentuin Buitenzorg,
Java: 375 (1892) (?); T.T.C.L.: 53 (1949); Pichon in Mém. I.F.A.N. 35: 85 (1953);
Persoon in Wageningen Agric. Univ. Pap. 92, 2: 202, t. 45 (1992); K.T.S.L.: 482
(1994). Type: Indonesia, Java, garden at Tjikeumeuh near Bogor Botanical Garden,
Schiffner 50 (L, holo., HBG, Z, iso.)

Slender liana 2–25 m high; branchlets dark reddish brown, glabrous. Leaves often
alternating with one or several pairs of bracts; blade ovate, elliptic to oblong, 2.5–12
cm long, 1.5–5.6 cm wide, acuminate at the apex, acumen rounded, 0.2–1.7 mm
long, rounded to acute at the base, coriaceous when fresh, glabrous on both sides;
secondary veins nearly straight, in 6–12 pairs; petiole 2–7 mm long, glabrous, with
colleters in the axils. Inflorescence predominantly axillary, but also terminal, 2–11-
flowered; peduncle 2–45 mm long, glabrous; pedicels 1–7 mm long, glabrous.
Flowers: sepals ovate or triangular, subequal, 0.9–1.5 mm long, ciliate; corolla white
or greenish white, glabrous outside; tube 2.6–4.8 mm long; lobes narrowly ovate,
3.2–7.4 mm long, rounded or obtuse at the apex. Fruit yellow or green, globose or
more or less pyriform, 1.4–5.5 cm long, 1.1–4.7 cm wide, with scattered lenticels 1–2
mm in diameter, 3–20 seeded; pericarp 0.7–7 mm thick; seeds irregularly ellipsoid or
ovoid, often laterally compressed, 6.8–14 mm long.

KENYA. Kwale District: Shimba Hills, Mkongani North, 4 May 1992, *Luke* 3124! & Gongoni
   Forest, 30 Dec. 1993, *PA & WRQ Luke* 3951B! & Mrima Hill, 5 Feb. 1989, *Mrima-Dzombo
   Expedition* 134!
TANZANIA. Lushoto District: Lutindi Forest Reserve, *Holst* 3443; Uzaramo District: Dar es
   Salaam, *Kirk* s.n.; Kilwa District: Selous Game Reserve, 10 km NNW of Kingupira, 20 Jan.
   1977, *Vollesen* 4361!
DISTR. **K** 7; **T** 3, 6, 8; **Z**; **P**; Mozambique
HAB. Moist forest; 0–250 m
USES. None recorded

NOTE. This species is cultivated in botanical gardens all over the world.

4. CLITANDRA

Benth. in Hook., Niger Fl. 445 (1849); Stapf in F.T.A. 4, 1: 60 (1902); Leeuwenberg
& Berndsen in B.J.B.B. 58: 159 (1988)

Lianas with tendrils; white latex present in all parts. Leaves opposite, inserted on
distinct leaf cushions; stipules absent. Inflorescence axillary, thyrsoid, congested or
lax. Flowers 5-merous, fragrant. Sepals green, without colleters. Corolla white,
pinkish or yellowish; corolla lobes overlapping to the left; stamens deeply included;
ovary 1–celled, with 2 parietal placentas, almost connate at the base, therefore often
seemingly 2–locular. Fruit a yellow or orange berry with black seeds embedded in
fibrous pulp; endosperm copious.

A monotypic African genus.

C. cymulosa Benth. in Hook., Niger Fl.: 445 (1849); Stapf in J.L.S. 30: 87 (1895) & in
F.T.A. 4, 1: 65 (1902); A. Chev. in Rev. Bot. Appliq. 28: 410 (1948); Pichon in Mém.
I.F.A.N. 35: 205 (1953); Huber in F.W.T.A. ed. 2, 2: 57 (1963); Leeuwenberg & Berndsen
in B.J.B.B. 58: 159, t. 1 (1988). Type: Sierra Leone, *G. Don* s.n. (K, holo., BM, iso.)

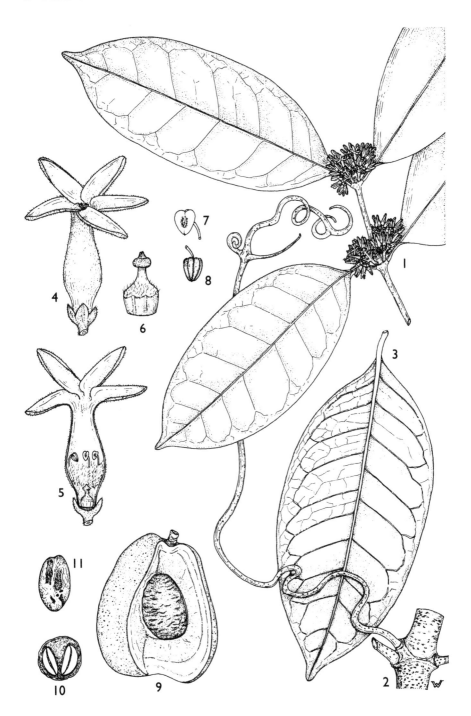

FIG. 7. *CLITANDRA CYMULOSA* — **1**, flowering branch, × ²/₃; **2**, branch with tendril, × ²/₃; **3**, leaf, × ²/₃; **4**, flower, × 6; **5**, opened corolla, × 6; **6**, gynoecium, × 18; **7–8**, stamen both sides, × 18; **9**, fruit, × ²/₃; **10**, CS fruit, × ¹/₃; **11**, seed, × ²/₃. 1 & 4–8 from *Hart* 282; 2 from *J.J. de Wilde* 8236; 3 from *Staner* 1341; 9–11 from *Louis* 5712. From B.J.B.B. vol. 58, drawn by W. Wessels, reproduced by permission.

Liana up to 40 m high; trunk up to 12 cm in diameter; bark rough, peeling; branches and branchlets lenticellate. Leaves petiolate, blade (narrowly) elliptic, 4.3–20 cm long, 1.8–8 cm wide, acuminate with the acumen obtuse and 1–15 mm long, cuneate at the base, glabrous on both sides or with scattered glandular hairs especially on the midrib and near the base; petiole 5–12 mm long, glabrous or with scattered glandular hairs. Inflorescence 1–2 together, 1.5–9.5 cm long, 10–80-flowered; peduncle 1–12 mm long, glabrous or pubescent; bracts 0.6–0.8 mm long. Flowers with sepals ovate, 0.6–1.3 mm long, 0.5–1 mm wide; corolla tube 3.1–6.4 mm long, widest at the insertion of the stamens, pilose outside, subglabrous above anthers; lobes (narrowly) oblong to (narrowly) ovate or obovate, 1.1–5.2 mm long, 0.5–1.8 mm wide, obtuse at the apex, entire, spreading; stamens inserted at 1–3 mm from the corolla base; anthers ovate, 0.3–0.5 mm long, apiculate, hanging or pointing downwards at anthesis; pistil 0.7–1.3 mm long, never touching stamens; style 0.2–0.4 mm long, glabrous; ovary 1–celled, with 6–15 ovules. Fruit ± 6.8 cm in diameter, globose or pyriform; exocarp with a hardened layer; seeds 6–25, 9–14 mm in diameter. Fig. 7 (p. 25).

UGANDA. Bunyoro District: Budongo Forest, *Dawe* 812; Mengo District: near Entebbe, Feb., *Chandler* 1573; Masaka District: Bugoma Port, Sese Is., 9 Dec. 1971, *Katende* 1432!
TANZANIA. Bukoba District: Bukoba, no date, *Stuhlmann* 1131!
DISTR. **U** 2–4; **T** 1; from Guinea to Congo (Kinshasa)
HAB. Forest or moist thicket margin; 1100–1200 m
USES. Fruit edible; latex used for high quality rubber

SYN. *Clitandra orientalis* K. Schum. in P.O.A. C: 315 (1895); Stapf in in F.T.A. 4, 1: 65 (1902); A. Chev. in Rev. Bot. Appliq. 28: 410 (1948); T.T.C.L.: 49 (1949). Type: Tanzania, Lake Province, Bukoba, *Stuhlmann* 1131 (B†, holo., K!, lecto., isolecto.)
   C. *gilletii* De Wild. in Obs. Apoc. Latex Rec. Congo (Brazzaville): 37 (1901). Type: Congo (Kinshasa), Bandundu, Iboko, *Gentil* 3 (BR, holo.)
   C. *arnoldiana* De Wild. in Comptes Rendus Acad. Sci., Paris 136: 400 (1903) & Miss. Laurent: 498 (1907). Type: Congo (Kinshasa), ancien district des Cataractes, *Serv. Agric.* 181 (BR, holo.)
   C. *nzunde* De Wild., Not. Apoc. Latic. Fl. Congo 1: 22 (1903); A. Chev. in Rev. Bot. Appliq. 28: 410 (1948). Type: Congo (Kinshasa), Equateur, 5 km S of Mobayi, *no collector* s.n. (BR, holo.)
   C. *arnoldiana* De Wild. var *seretii* De Wild., Miss. Laurent: 500 (1907). Type: Congo (Kinshasa), upper Congo R., Arebi, *Seret* 730 (BR, holo.)

5. DICTYOPHLEBA

Pierre in Bull. Soc. Linn. Paris ser. 2, 1: 92 (1898); Pichon in Mém. I.F.A.N. 35: 250 (1953); de Hoogh in B.J.B.B. 59: 207–226 (1989)

Landolphia P. Beauv. sect. *Dictyophleba* (Pierre) Hallier f. in Jahrb. Hamb. Wiss. Anst. 17, Beih. 3: 69 (1900)

Lianas with curled tendrils; white, sticky latex present in all parts; branches with conspicuous transverse leaf scars, lenticellate, virgate-sarmentose. Leaves opposite, exstipulate. Inflorescence a loose, elongate, terminal panicle of cymes; pedunculate, often tendril-like. Corolla glabrous outside, inside pilose at the widest part of the almost cylindrical tube; corolla lobes overlapping to the left, long-ciliate along the margin that is covered in bud; stamens inserted at the widened part of the tube; anthers cordate at the base, glabrous; ovary of two united carpels, disk absent; pistil head composed of an oblong stigmatic basal part and a cleft stigmoid apex. Fruit a red, orange or yellow berry; pulp yellow. Seeds brown; cotyledons flat, undulate.

A genus of 5 species confined to Africa.

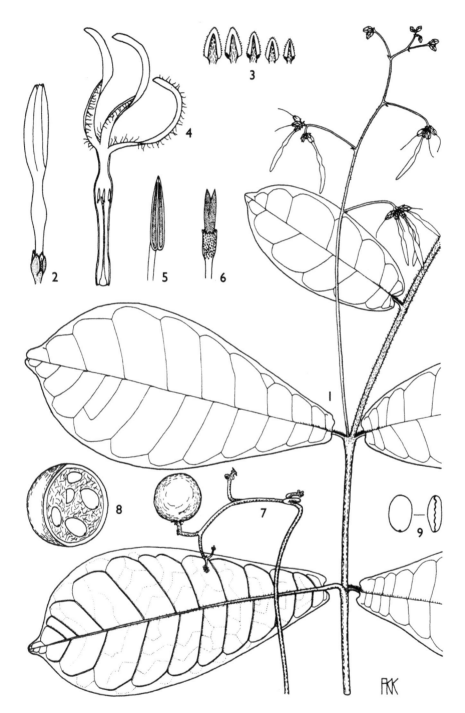

FIG. 8. *DICTYOPHLEBA LUCIDA* — **1**, habit, × ²/₃; **2**, flower bud, × 2; **3**, sepals, × 4; **4**, section of flower, × 2; **5**, stamen, × 12; **6**, apex of gynoecium, × 12; **7**, fruiting shoot, × ²/₃; **8**, section of fruit, schematic, × ²/₃; **9**, seed, dorsal and lateral views, × 1. 1 from *Pawek* 6418 and *Garcia* 296; 3 from *Louis* 3737; 2, 4–6 from *Barbosa* 1252; 7–9 from *Torre & Correia* 1437. From F.Z., drawn by F.K. Kupicha.

D. lucida (*K. Schum.*) *Pierre* in Bull. Soc. Linn. Paris ser. 2, 1: 93 (1898); Pichon in Mém. I.F.A.N. 35: 262, t. 3 & 4 (1953); Huber in F.W.T.A. ed. 2, 2: 52 (1963); Kupicha in F.Z. 7 (2): 420, t. 95 (1985); de Hoogh in B.J.B.B. 59: 212, t. 3 (1989); K.T.S.L.: 479, ill. (1994). Type: Congo (Kinshasa), Mukenge, *Pogge* 1038 (B†, holo., K!, lecto.)

Liana 1–20(–35) m high; trunk 0.6–8 cm in diameter; bark pale greenish brown; branches dark brown, glabrous; branchlets glabrous or hirsute. Leaves petiolate, stipulate; blade obovate or elliptic, 3.2–17 cm long, 1.6–8 cm wide, bluntly acuminate to retuse at the apex, cordate or subcordate at the base, with 6–13 pairs of distinct, strongly looped secondary veins forming a submarginal vein; stipules interpetiolar, 1–1.5 mm long, early caducous, with colleters in axils; petiole 3–13 mm long, pubescent or hirsute. Inflorescence up to 23 cm long, cymes alternate, 2–20-flowered. Flowers slightly scented; sepals almost free, ovate to oblong, 1.5–2.9 mm long, minutely ciliate; corolla pale yellow or pinkish in bud, becoming white at anthesis, often with a pink to red tube; tube 10–18 mm long; lobes narrowly oblong, 7.5–20 mm long, 1.5–2.8 mm wide, hairs on inner margin 2–3 mm long; stamens inserted at (5–)7.5–10.5 mm from the base; anthers narrowly oblong, 1.7–2.1 mm long; pistil 7.2–13 mm long; style 5.5–10 mm long; pistil head 0.4–0.6 mm long, stigmoid apex 0.8–1.2 mm long. Fruit red, orange or yellow, subglobose, 12–50 mm long, 10–35 mm in diameter, 1–29-seeded; pulp yellow, acidic. Fig. 8 (p. 27).

KENYA. Kilifi District: Kaya Jibana, 14 Dec. 1990, *Luke & Robertson* 2651!; Kwale District: Shimba Hills, Longomwagandi Forest, 2 June 1968, *Magogo & Glover* 1150! & Buda Mafisini Forest, 21 Aug. 1953, *Drummond & Hemsley* 1941!
TANZANIA. Kigoma District: Kaskati Basin, 80 km S of Kigoma, 5 Mar. 1964, *Itani* 155!; Morogoro District: Lusunguri Forest Reserve, 31 Mar. 1953, *Drummond & Hemsley* 1941!; Lindi District: Rondo Plateau, St. Cyprians College, 18 Feb. 1991, *Bidgood et al* 1630!
DISTR. **K** 7; **T** 3, 4, 6–8; Nigeria, Cameroon, Central African Republic, Congo (Kinshasa), Burundi, Angola, Zambia, Mozambique, Malawi, Zimbabwe and Comoro Islands
HAB. Forest margins, riverine or gallery forest; 0–950 m
USES. Fruit pulp edible

SYN. *Landolphia lucida* K. Schum. in N.B.G.B. 1: 24 (1895) & in E. & P. Pf. 4, 2: (1895); T.T.C.L.: 51 (1949)
　　L. lucida K. Schum. var. *hispida* Hallier f. in Jahrb. Hamb. Wiss. Anst. 17, Beih. 3: 86 (1900); T.T.C.L.: 51 (1949). Type: Tanzania, Lushoto District, Usambara, *Scheffler* 217 (B†, holo., BM, lecto.; E, EA!, K, P, Z, iso., chosen by de Hoogh)
　　L. debreucquiana De Wild. in De Wild. & Gentil, Lianes Caoutch. Et. Congo: 92 (1904). Lectotype: Congo (Kinshasa), without locality, *Dewèvre* 1036a (BR, lecto., chosen by de Hoogh)

6. ANCYLOBOTRYS

Pierre in Bull. Soc. Linn. Paris sér. 2: 91 (1898); Pichon in Mém. I.F.A.N. 35: 272 (1953) (both as *Ancylobothrys*); Vonk, Leeuwenberg & Haegens in Wageningen Agric. Univ. Pap. 94 (3): 4–44 (1994)

Sarmentose shrubs or lianas, at least partly with a rusty brown indumentum, with large curled tendrils; latex present. Leaves opposite, petiolate, exstipulate. Inflorescence a long terminal panicle, with sensitive branches, these later becoming climbing hooks. Flowers sweet-scented. Sepals green, often seemingly rusty brown by indumentum, usually without colleters. Corolla white or nearly so; tube narrow, often thickened above the stamens, glabrous or pubescent outside; corolla lobes overlapping to the left, fringed with white hairs; stamens deeply included, inserted at the lower quarter of the tube; anthers narrowly triangular, cordate at the base, keeled along the back, glabrous; ovary unilocular, hairy, abruptly narrowed into the style; pistil head of a subglobose or cylindrical stigmatic basal part and a short, bilobed stigmoid apex. Fruit a berry, yellow to red, globose or pear-shaped or nearly so, rounded at the apex, softly velutinous. Seed with horny endosperm, surrounding the embryo.

A genus of 7 species in tropical Africa with *A. petersiana* extending to the Comoro Islands and NW Madagascar.

1. Secondary veins in 19–40 pairs, 1–4 mm apart; branchlets
   glabrous or minutely dark brown pubescent · · · · · · · · · ·   3. *A. tayloris*
   Secondary veins in 4–14 pairs, 5–20 mm apart; branchlets
   almost tomentose or rusty brown pubescent · 2
2. Leaf apex acuminate, secondary veins in 7–14 pairs;
   inflorescence with 5–11 cymes; fruits 15–20 mm in
   diameter ·   1. *A. amoena*
   Leaf apex rounded or acute; secondary veins in 4–9 pairs;
   inflorescence with 2–6 cymes; fruits 25–50 mm in diameter ·   2. *A. petersiana*

1. **A. amoena** *Hua* in Bull. Mus. Nat. Hist. Nat. Paris 5: 186 (1899); Pichon in Mém. I.F.A.N. 35: 274, t 17/1–2 (1953); Huber in F.W.T.A. ed. 2, 2: 60 (1963); Kupicha in F.Z. 7 (2): 426 (1985); Vonk et al. in Wageningen Agric. Univ. Pap. 94 (3): 5, t. 1 (1994). Type: Guinea, Kankan, Upper Niger R. near Kouroussa, *Paroisse* 24 (P, holo., K, iso.)

Scrambling shrub or liana up to 10 m high; branchlets densely brown-pubescent to almost tomentose. Leaves petiolate; blade elliptic or ovate, rarely oblong, 6.7–12 cm long, 2.8–7 cm wide, apex acuminate with acumen up to 10 mm long, cuneate at the base or decurrent into the petiole, glabrous to minutely pubescent on both sides, with 7–14 pairs of conspicuous straight to slightly curved secondary veins, forming an angle of 45–70° with the midrib; petiole densely pubescent, 5–13 mm long. Inflorescence 10–38 cm long, with 5–11 cymes, each with about 25 flowers, densely pubescent in all parts; peduncle 4–18 cm long; bracts sepal-like, 1.7–2 mm long; pedicels 1 mm long. Flowers with sepals triangular to ovate, 2.5–5 mm long, ciliate; corolla white or pinkish, especially the tube; tube 11–18 mm long; lobes narrowly elliptic, rarely oblong, 8.5–21 mm long, obtuse at the apex; stamens inserted at 2.7–3.9 mm from the base; pistil 3.5–4.5 mm long; ovary subglobose, ovules 40–50; style 1.5–2.1 mm long, glabrous. Fruit yellow to red, globose or slightly pear-shaped, 1.5–2 cm in diameter, 1–4 seeded; seeds 7–10 mm long.

UGANDA. W Nile District: Koboko, *Eggeling* 1838; sine loc., *Bagshawe* 205
TANZANIA. Mwanza District: Rubya Forest Reserve, 2 Feb. 1960, *Carmichael* 745!; Mpanda District: Mahali Mts, 18 Apr. 1978, *Uehara* 535; Bagamoyo District: Bana Forest Reserve, 5 Aug. 1968, *Shabani* 108!
DISTR. U 1; T 1, 4, 5, 7, 8; W Africa from Guinea and Mali to Chad and from Sudan to Zambia
HAB. Woodland, bushland and thicket, often on rocky hills; 750–1650 m
USES. Fruit edible

SYN. *Landolphia scandens* (K. Schum. & Thonn.) Didr. var. *ferruginea* Hallier f. in Jahrb. Hamb. Wiss. Anst. 170, Beih. (3): 80 (1900). Type: Tanzania, Lake Malawi, Kanda Peninsula, *Goetze* 884 (K, lecto, BR, E, isolect., chosen by Vonk et al.)
　　L. scandens (K. Schum. & Thonn.) Didr. var. *rigida* Hallier f. in Jahrb. Hamb. Wiss. Anst. 170, Beih. (3): 81 (1900). Lectotype: Tanzania, Bukoba District, Karagwe, *Scott Elliot* 8167 (K, lecto., chosen by Vonk et al.)
　　L. scandens (K. Schum. & Thonn.) Didr. var. *schweinfurthiana* Hallier f. in Jahrb. Hamb. Wiss. Anst. 170, Beih. (3): 81 (1900). Lectotype: Sudan, Djurland, Zeriba Agat on Wau R., *Schweinfurth* 1685 (K, lecto., P, iso., chosen by Vonk et al.)
　　L. amoena (Hua) Hua & A. Chev. in J. Bot. (Morot) 15: 8 (1901)
　　L. ferruginea (Hallier f.) Stapf in F.T.A. 4, 1: 46 (1902)
　　L. petersiana (Klotzsch) Dyer var. *schweinfurthiana* (Hallier f.) Stapf in F.T.A. 4, 1: 48 (1902); T.T.C.L.: 51 (1949)
　　Pacouria petersiana (Klotzsch) S. Moore var. *schweinfurthiana* (Hallier f.) S. Moore in J.L.S. 37: 180 (1905)
　　P. amoena (Hua) Pichon in Mém. Mus. Nat. Hist. Nat., Paris sér. 2 (24): 144 (1948)
　　Landolphia nitida Lebrun & Taton, Expl. Parc Nat. Kagera: 105 (1948). Type: Rwanda, Ryabega [Ruabiega], *Lebrun* 9820 (BR, holo., K, P, iso.)

2. **A. petersiana** (*Klotzsch*) *Pierre* in Bull. Soc. Linn. Paris, sér. 2: 91 (1898); Pichon in Mém. I.F.A.N. 35: 290, t. 21/1–3 (1953); Kupicha in F.Z. 7 (2): 423 (1985); K.T.S.L.: 477, ill., map (1994); Vonk et al. in Wageningen Agric. Univ. Pap. 94 (3): 13, t. 3 (1994). Type: Mozambique, Sofala, Sena, *Peters* s.n. (B†, holo.); Niassa, Angoche [Antonio Enes], *Gomes & Sousa* 4856 (K, neo., COI, PRE, WAG, isoneo., designated by Vonk 1994)

Scrambling shrub or liana up to 5 m high; trunk 2–5 cm in diameter; branches lenticellate; branchlets densely rusty brown pubescent. Leaves petiolate; blade elliptic or slightly obovate, 4.5–9.3 cm long, 2.2–4.4 cm wide, rounded or acute at the apex, never with a clear acumen, cuneate or rounded at the base or decurrent into the petiole, coriaceous when dried, glabrous to minutely pubescent on both sides, especially on the veins; with 4–9 pairs of curved secondary veins, at an angle of 45–60° with the midrib; petiole 6–10 mm long, glabrous to pubescent. Inflorescence 7.5–37 cm long, with 2–6 cymes, each with ± 20 flowers; peduncle 4.2–25 cm long; bracts sepal-like, 1.4–2.2 mm long; pedicels 1–2 mm long. Flowers with sepals triangular to ovate, 1.7–3.9 mm long; corolla white or cream, often pinkish at the tube; tube narrow, 8.5–13 mm long; lobes narrowly elliptic to oblong, 11–35 mm long, 2–4.5 mm wide, glabrous; stamens inserted at 1.8–3 mm from the base of the tube; pistil 2.2–3.8 mm long; ovary subglobose, ovules 22–50. Fruit yellow or orange, globose to slightly pear-shaped, 25–50 mm in diameter, about 5–20 seeded; seeds ovoid or ellipsoid, up to 18 mm long. Fig. 9/7–8 (p. 31).

KENYA. Kilifi District: Arabuko Sokoke Forest, 16 Sept. 1985, *Beentje* 2286!; Lamu District: SW edge of Boni Reserve, 19 Oct. 1980, *Kuchar* 13531!; Kwale District: Shimba Hills, 15 Nov. 1968, *Magogo & Estes* 1241!

TANZANIA. Lushoto District: W Usambara, 1939, *Gillman* 878!; Morogoro District: 12.5 km NE of Kingolwira, 30 Nov. 1957, *Welch* 443!; Uzaramo District: Sinda Island, 5 Jan. 1969, *Harris* 2703!

DISTR. **K** 1/7, 7; **T** 1–4, 6–8, **P**, **Z**; Congo (Kinshasa), Burundi, Somalia, Malawi, Mozambique, Zimbambwe, South Africa, Comoro Islands and Madagascar

HAB. Light forest, bushland and coastal thicket; 0–600(–1150) m

USES. Roots and leaves used in minor medicines

SYN. *Willughbeia petersiana* Klotzsch in Peters, Reise Mossamb. Bot. 1: 281 (1862)
> *W. sennensis* Klotzsch in Peters, Reise Mossamb. Bot. 1: 281 (1862). Type: Mozambique, Sofala, Sena, *Peters* s.n. (B†, holo.)
> *Landolphia petersiana* (Klotzsch) Dyer in Report. Roy. Bot. Gard. Kew 1880: 42 (1881); U.O.P.Z.: 324 (1949); T.T.C.L.: 52 (1949); Codd in F.S.A. 26: 260 (1963)
> *L. monteiroi* N. E. Br. in Monteiro, Delagoa Bay: 161, 163, 178 (1891); Stapf in K.B. 1907: 51 (1907). Type: Mozambique, Delagoa Bay, *Monteiro* 37 (K, holo.)
> *L. angustifolia* Engl. in Abh. Kön. Akad. Wiss. Berlin, Phys.-Math 1: 34 (1894); K. Schum. in N.B.G.B. 1: 25 (1895). Lectotype: Tanzania, Tanga District, Misoswe [Mizozue], *Holst* 2220 (B†, holo., K, lecto., COI, HBG, M, isolecto., chosen by Vonk et al.)
> *L. petersiana* (Klotzsch) Dyer var. *rotundifolia* Dew., Ann. Soc. Sci. Brux. 19 (II): 122 (1895). Type: Tanzania, Zanzibar, *Joblonsky* 23 (P, holo.)
> *L. sennensis* (Klotzsch) K. Schum. in P.O.A. B: 453 (1895)
> *Ancylobotrys rotundifolia* (Dew.) Pierre in Bull. Soc. Linn. Paris. sér. 2: 92 (1898)
> *Landolphia scandens* (K. Schum. & Thonn.) Didr. var. *angustifolia* (Engl.) Hallier f. in Jahrb. Hamb. Wiss. Anst. 17, Beih. (3): 84 (1900)
> *L. scandens* (K. Schum. & Thonn.) Didr. var. *petersiana* (Klotzsch) Hallier f., in Jahrb. Hamb. Wiss. Anst. 17, Beih. (3): 82 (1900)
> *L. scandens* (K. Schum. & Thonn.) Didr. var. *rotundifolia* (Dew.) Hallier f. in Jahrb. Hamb. Wiss. Anst. 17, Beih. (3): 82 (1900)
> *L. scandens* (K. Schum. & Thonn.) Didr. var. *stuhlmanii* Hallier f. in Jahrb. Hamb. Wiss. Anst. 17, Beih. (3): 83 (1900). Type: Tanzania, Tanga District, Amboni, *Holst* 2563 (B, holo †, HBG, lecto., COI, K, M, W, Z, isolecto., chosen by Vonk et al.)
> *L. petersiana* (Klotzsch) Dyer var. *angustifolia* (Engl.) Stapf in F.T.A. 4, 1: 48 (1902); T.T.C.L.: 52 (1949)
> *L. petersiana* (Klotzsch) Dyer var. *rufa* Stapf in F.T.A. 4, 1: 48 (1902). Type: Malawi, sine loc., Buchanan 437 (K, lecto)

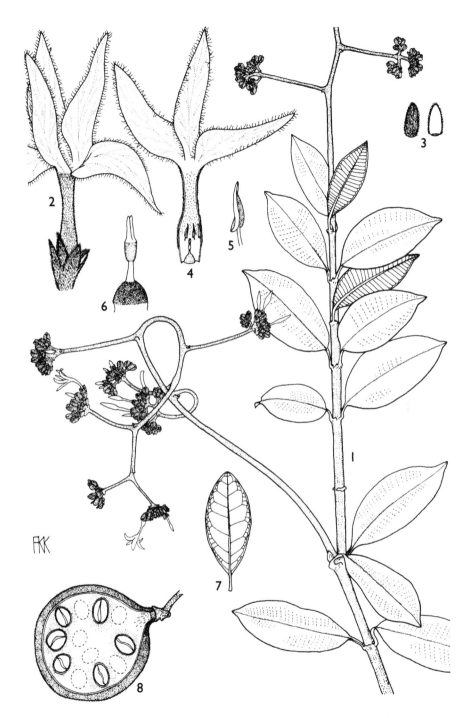

FIG. 9. *ANCYLOBOTRYS TAYLORIS* — **1**, habit, × ²/₃; **2**, flower, × 2¹/₂; **3**, sepals, inside and outside, × 2¹/₂; **4**, longitudinal section of flower, × 2¹/₂; **5**, stamen, × 10; **6**, gynoecium, × 10. *ANCYLOBOTRYS PETERSIANA* — **7**, leaf, × ²/₃; **8**, cross-section of fruit, schematic, × ²/₃. 1–6 from *Mendonca* 959; 7 from *Andrada* 1917, 8 from *Torre & Correia* 17032. From F.Z., drawn by F.K. Kupicha.

L. petersiana (Klotzsch) Dyer var. *tubeufii* Stapf in F.T.A. 4, 1: 49 (1902). Lectotype: Tanzania, sine loc., *Busse* 1051 (G, lecto, B, HBG, isolecto., chosen by Vonk et al.)
L. scandens (K. Schum. & Thonn.) Didr. var. *tubeufii* (Stapf) Busse in E.J. 32: 171 (1902)
Pacouria angustifolia (Engl.) O. Kuntze in Deutsche Bot. Monatss. 21: 173 (1903)
P. petersiana (Klotzsch) S. Moore in J.L.S. 37: 180 (1905)

3. **A. tayloris** (*Stapf*) Pichon in Mém. I.F.A.N. 35: 284, t. 18/1–3 (1953); Kupicha in F.Z. 7, 2: 423, t. 106 (1985); K.T.S.L.: 477, map (1994); Vonk et al. in Wageningen Agric. Univ. Pap. 94 (3): 30, t. 7 (1994). Type: Kenya, Kilifi District, Rabai Hills, Fimboni, *W. E. Taylor* s.n., Nov. 1885 (BM, holo.)

A climbing shrub or liana up to 20 m high; branches dark brown, lenticellate; branchlets dark brown pubescent or sometimes glabrous. Leaves petiolate; blade elliptic, 4–12 cm, long, 1.7–5 cm wide, apex acuminate with acumen up to 17 mm long, cuneate at the base, glabrous on both sides, with 19–40 pairs of more or less straight conspicuous secondary veins at an angle of 60–90° with the midrib; petiole 5–8 mm long, glabrous to puberulous, with colleters in the axils; veins 1–4 mm apart. Inflorescence 9–32 cm long, with 5–9 cymes, each with about 30 flowers, pubescent in all parts; peduncle 4–12.5 cm long; bracts up to 2.5 cm long; pedicels 1 mm long. Flowers with sepals triangular to ovate, 1.6–3.1 mm long; corolla white; tube pale brown, 8–12.5 mm long; lobes narrowly triangular to narrowly elliptic, 10–20 mm long, 2–5 mm wide, glabrous on both sides; stamens inserted at 2–2.9 mm from the base of the tube; pistil 2.5–2.9 mm long; ovary subglobose to slightly cylindrical, ovules 15–35. Fruit yellow, subglobose, to 4 cm in diameter. Fig. 9/1–6 (p. 31)

KENYA. Kwale District: Buda Forest, 21 Aug. 1953, *Drummond & Hemsley* 3935!; Kilifi District: Arabuko, Aug. 1965, *Langridge* 79! & Kaya Kambe, 10 Mar. 1981, *Hawthorne* 83C!
TANZANIA. Kilwa District: Balenje, Nov. 1968, *Rodgers* 521!; Newala District: Newala, *Hay* 73; Uzaramo District: Pande Forest Reserve, July 1989, *Rulangaranga et al.* 73
DISTR. **K** 7; **T** 6, 8; Malawi and Mozambique
HAB. Coastal forest; 0–700 m
USES. Fruit edible

SYN. *Landolphia tayloris* Stapf in F.T.A. 4, 1: 45 (1902)
    Landolphia pachyphylla Stapf in F.T.A. 4, 1: 45 (1902). Type: Malawi, sine loc., *Buchanan* 140 (BM, holo., E, iso.)

7. **SABA**

(Pichon) Pichon in Mém. I.F.A.N. 35: 302 (1953); Leeuwenberg & van Dilst in B.J.B.B. 59: 189–206 (1989)

Landolphia sect. *Saba* Pichon in Mem. Mus. Nat. Hist. Nat., Paris, nov. ser. 24: 140 (1948)

Lianas with curled tendrils and white latex in all parts. Leaves opposite, with some colleters in axils. Inflorescences terminal in forks or on the tendrils. Flowers sweet-scented. Corolla white or creamy with orange or yellow throat; tube inside with a pilose belt around the anther; lobes overlapping to the left; stamens inserted below the middle of the tube; anthers glabrous, acute at the apex, cordate at the base; ovary glabrous or with a setose belt just below the truncate apex; pistil head oblong or ellipsoid. Fruit a many-seeded berry with a glabrous, smooth or rough sclerified pericarp and edible pulp.

A genus of three species in Tropical Africa.

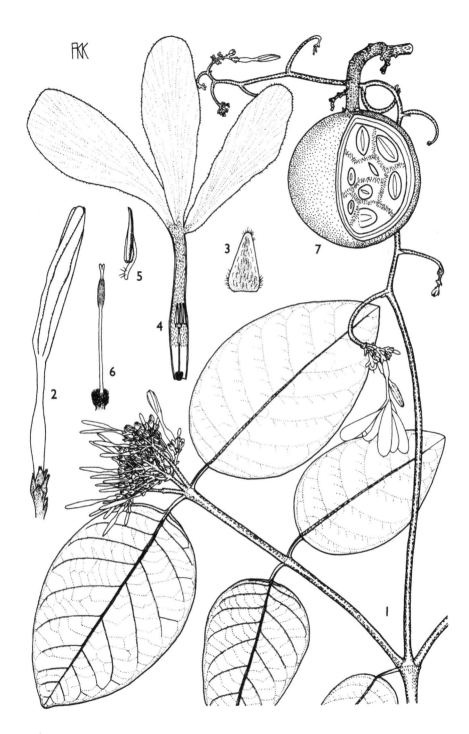

FIG. 10. *SABA COMORENSIS* — **1**, habit, × ²/₃; **2**, flower bud, × 2; **3**, sepal, × 6; **4**, section of flower, × 2; **5**, stamen, × 6; **6**, gynoecium, × 6; **7**, fruit, × ²/₃. 1 from *Garcia* 931, 2–6 from *Torre & Paiva* 10615, 7 from *Torre & Correia* 15865. From F.Z., drawn by F.K. Kupicha.

S. comorensis (*Bojer*) *Pichon* in Mém. I.F.A.N. 35: 303 (1953); Kupicha in F.Z. 7 (2): 427 (1985); Wild Flow. E. Afr.: 143, t. 70 (1987); Leeuwenberg & van Dilst in B.J.B.B. 59: 190, t. 1–3 (1989); U.K.W.F. ed. 2: 171, ill. (1994); K.T.S.L.: 484, ill., map (1994). Type: Comoro Islands, *Bojer* s.n. (K, iso.)

Liana up to 20 m long; trunk up to 15 cm in diameter; wood reddish or yellow; bark rough, scaly, often fissured. Leaves petiolate; blade ovate or elliptic, 4.5–24 cm long, 3.4–12 cm wide, rounded, obtuse or rarely shortly acuminate at the apex, oblique, rounded or subcordate at the base; petiole 7.5–20 mm long. Inflorescence a many-flowered congested cyme; peduncle 2–6(–32) cm long; pedicels 1–7 mm long. Flowers white with a yellow or orange throat; sepals 1.7–4 mm long, 1–2 mm wide; tube greenish, 16–34 mm long; lobes 19–41 mm long, 4–12 mm wide; ovary with a ring of white stiff hairs at the apex, ovules ± 50; stigmoid apex up to 2.2 mm long. Fruit yellow or orange, (sub-)globose, up to 11 cm in diameter; pericarp 1–2.4 cm thick; very young fruits hairy; seeds whitish, yellow to brown, up to 1 cm in diameter. Fig. 10 (p. 33).

UGANDA. Bunyoro District: Budongo Forest, Mar. 1973, *Synnott* 1441; Teso District: Serere, Sep. 1932, *Chandler* 942; Busoga District: Lolui Island, May 1964, *G. Jackson* 12
KENYA. Embu District: Siakago, near Consolata mission, 20 Aug. (year?), *Riley* 74047!; Kisumu–Londiani District: Escarpment near Kisumu, Sept. 1958, *Tweedie* 1702!; Teita District: Taita Hills, Mbololo Hill, 14 Feb. 1953. *Bally* 8580!
TANZANIA. Mwanza District: Rubya Forest Reserve, Ukerewe Is., 2 Mar. 1902, *Carmichael* 862!; Kigoma District: Kaskati, Feb. 1965, *Suzuki* 45!; Uzaramo District: Pugu Hills, 19 Oct. 1969, *Harris et al.,* 3512!
DISTR. **U** 1–4; **K** 1–7; **T** 1–4, 6–8; **P**; from Senegal to Madagascar and Ethiopia to Zimbabwe, Comoro Is.
HAB. Moist forest, forest edges, secondary forest, thicket, woodland in rocky sites; 0–2000 m
USES. Wood for walking sticks; latex for rubber; fruit edible, sold in some coastal markets, and also much appreciated by monkeys; minor medicinal against snakebite, worms

SYN. *Vahea comorensis* Bojer in Hort. Maurit. 207 (1837); A. DC. in Prodr. 8: 328 (1844)
   Landolphia florida Benth. in Niger Fl.: 444 (1849); Stapf in F.T.A. 4, 1: 38 (1902); U.O.P.Z.: 323 (1949). Type: Nigeria, Quorra, along the banks of the Niger R., *Vogel* 101 (K, holo.)
   Willughbeia cordata Klotzsch in Peters, Reise Mossamb.1: 283 (1875). Type: Comoro Is., Anjouan, *Peters* s.n. (B†, holo.)
   Landolphia florida Benth. var. *leiantha* Oliv. in Trans. Linn. Soc. London 29: 107 (1875); Stapf in F.T.A. 4, 1: 39 (1902). Type: Uganda, W Nile District, Madi, *Speke & Grant* 707 (K, holo.)
   L. comorensis (Bojer) K. Schum. in E.J. 15: 406 (1895); T.T.C.L.: 51 (1949)
   L. comorensis (Bojer) K. Schum. var. *florida* (Benth.) K. Schum. in E.J. 15: 404, t. 1b & 2 (1895)
   Saba comorensis (Bojer) Pichon var. *florida* (Benth.) Pichon in Mém. I.F.A.N. 35: 309 (1953)
   S. florida (Benth.) Bullock in K.B. 13: 391 (1959); F.P.U.: 118, t. 66 (1962); Huber in F.W.T.A. ed. 2, 2: 61 (1963); Drummond in Kirkia 10: 269 (1975)

8. PICRALIMA

Pierre in Bull. Soc. Linn. Paris 2: 1278 (1896); Stapf in F.T.A. 4, 1: 96 (1902); Omino in Wageningen Agric. Univ. Pap. 96 (I): 128–134 (1996).

Tree or shrub, glabrous in all parts except inside the corolla tube; white latex present in all parts. Leaves opposite. Inflorescence terminal, rarely axillary, a compound umbellate cyme, pedunculate. Flowers fragrant or not. Sepals imbricate, with 2–4 rows of colleters at the extreme base. Corolla tube thick and fleshy, almost cylindrical; lobes overlapping to the left, entire, spreading and later recurved; stamens included; filaments short; ovary of two separate carpels, united at the extreme base by a disk-like thickening; ovules numerous in each carpel; pistil head of a stigmatic oblong basal part and a stigmoid apex up to 1.5 mm long. Fruit of two separate mericarps, up to 80-seeded.

Monotypic and African.

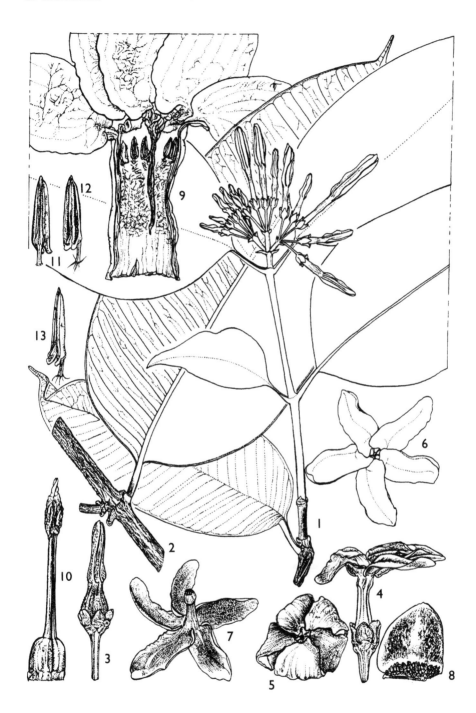

FIG. 11. *PICRALIMA NITIDA* — **1**, habit, × ²/₃; **2**, part of branch, × ²/₃; **3**, flower bud, × 1; **4**, flower, × 1; **5–6**, flower from above, × 1; **7**, corolla from underside, × 1; **8**, sepal inside, × 2; **9**, opened corolla, × 2; **10**, pistil, × 4¹/₂; **11–13**, stamens, × 6¹/₂. 1–2 from *W. de Wilde et al.* 1952; 3–13 from *Reitsma* 899. From Meded. Landbouwhogeschool: 96(1), drawn by ?, reproduced by permission.

P. nitida (*Stapf*) *Th. & H. Durand* in B.J.B.B. 2: 338 (1910); I.T.U. ed. 2: 28 (1952); Huber in F.W.T.A. ed. 2, 2: 62 (1963); Omino in Wageningen Agric. Univ. Pap. 96 (1): 128, t. 35 (1996). Type: Cameroon, Ambas Bay, *Mann* 710 (K!, holo., GH, P!, iso.)

Tree or shrub 4–35 m high, crown dense; trunk 5–60 cm in diameter, cylindrical; wood very hard. Leaves petiolate; blade broadly to narrowly ellipic to oblong, 10–27 cm long, 1.9–13 cm wide, apex abruptly acuminate with acumen 6–14 mm long, rounded to obtuse at the base; petiole 10–22 mm long. Inflorescence 6–10 cm long, 12–36-flowered; peduncle 2–37 mm long; bracts small; pedicels 2–20 mm long. Flowers with sepals broadly ovate to suborbicular; 3.5–7.5 mm long, 3.5–5 mm wide, obtuse to rounded at the apex; corolla tube often greenish, 14.5–21 mm long; lobes white to yellow, ovate, 13.5–30 mm long, 5.5–10 mm wide, obtuse at the apex; stamens inserted at 9.5–15 mm from the base; pistil 9.7–12.5 mm long; style 5–7.3 mm long; ovules 120–150 in each carpel. Fruits yellow to orange, smooth, irregularly pyriform or obovoid to ellipsoid, 8–15 cm long, 13–14 cm in diameter, apex rounded. Fig. 11 (p. 35).

UGANDA. Toro District: Semliki Valley, 28 Nov. 1905, *Dawe* 719!; Bunyoro District: Bugoma, 28 Nov, 1905, *Dawe* 707! & Budongo Forest, Aug. 1940, *Eggeling* 4042!
DISTR. U 2; Ivory Coast, Ghana, Nigeria, Cameroon, Gabon, Congo (Brazzaville) and Congo (Kinshasa)
HAB. Rainforest; 800–1200 m
USES. Bark chewed as a vermifuge

SYN. *Tabernaemontana nitida* Stapf in K.B. 1894: 22 (1894)

9. HUNTERIA

Roxb. in Fl. Ind. Seramp. ed. Carey & Wall. 2: 531 (1824) & ed. 2 (1): 695 (1832); Omino in Wageningen Agric. Univ. Pap. 96 (1): 88–128 (1996)

Shrubs or trees, rarely lianas, glabrous in all parts except inside the corolla tube; white or colourless latex present in all parts. Leaves opposite. Inflorescence mostly terminal, sometimes also axillary, few–many-flowered cymes, dense to lax, pedunculate. Flowers fragrant. Sepals free or connate, with colleters covering about half of the sepal length. Corolla tube much longer than calyx, almost cylindrical; corolla lobes overlapping to the left in bud, sometimes twisted, spreading and recurved later; stamens included; ovary of 2 separate carpels united at the extreme base by a disk-like thickening; pistil head of a stigmatic subglobose basal part and a stigmoid apex up to 1 mm long; ovules 1–6(–30) in each carpel. Fruit smooth or warty (not in our area), of two separate mericarps, with a rounded to beaked apex. Seeds 1–9(–26), brown, variously shaped.

12 species in Africa; *H. zeylanica* extends to S and SE Asia.

Corolla tube 4.8–5.5 mm long; ovary gradually narrowing into
    the style; ovules 5–6 in each carpel · · · · · · · · · · · · · · · · · 1. *H. congolana*
Corolla tube 6–10 mm long; ovary abruptly narrowing into the
    style; ovules 2 in each carpel · · · · · · · · · · · · · · · · · · · 2. *H. zeylanica*

1. **H. congolana** *Pichon* in Bol. Soc. Brot. sér. 2 (27): 101, t. 1/9–13 (1953); K.T.S.L.: 480 (1994); Omino in Wageningen Agric. Univ. Pap. 96 (1): 96, t. 25 (1996). Type: Congo (Kinshasa), 10 km E of Yangambi, *J. Louis* 1083 (BR!, holo., K!, P, S, iso.)

Shrub or tree 1.7–20 m high; latex white to yellow in all parts; trunk 5–30 cm in diameter, wood yellow and hard. Leaves petiolate; blade elliptic to oblong, 6.3–18.5 cm long, 1.5–5.9 cm wide, acute to obtuse at the base, apex acuminate with acumen

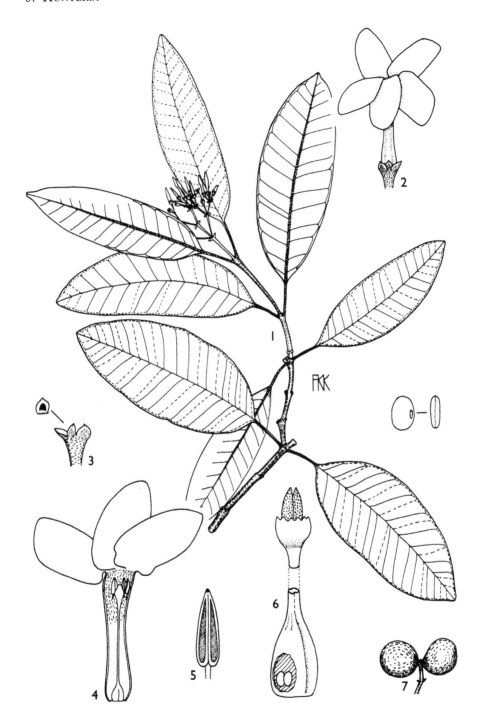

FIG. 12. *HUNTERIA ZEYLANICA* — **1**, habit × ²⁄₃; **2**, flower, × 2¹⁄₂; **3**, part of calyx,× 4; **4**, part of corolla opened out, × 4; **5**, stamen, × 20; **6**, apex and base of gynoecium, part of wall removed, × 20. 1 from *Andrada* 1062; 2–6 from *Tinley* 2369; 7–8 from *Müller & Pope* 1865. From F.Z., drawn by F.K. Kupicha.

10–18 mm long; petiole 8–18 mm long. Inflorescence terminal or axillary, 3–3.5 cm long, 7–18(–45)-flowered; peduncle 2–14 mm long; bracts up to 1.5 mm long, with colleters within; pedicels 2–5 mm long. Flowers with sepals with resin within, (narrowly) ovate, 0.8–1.5 mm long; corolla white or yellow, creamy pink in bud; tube 4.8–5.5 mm long; lobes narrowly ovate to oblong, 4–8 mm long, 1.5–2.8 mm wide, obtuse; stamens inserted at 3–4.7 mm from the base; pistil 3.4–4.8 mm long; ovary gradually narrowing into the style, ovules 5–6 in each carpel; style 0.8–2.1 mm long. Fruits yellow to orange, ellipsoid to subglobose, 20–26 mm long, 11–18 mm in diameter, 1–3-seeded.

KENYA. Northern Frontier District: Mt Kulal, *Adamson* 115! & N of Gatab, *sin. coll* 178!
DISTR. **K** 1; Congo (Kinshasa)
HAB. Forest; 1700–2100 m
USES. None recorded

NOTE. The main area of distribution is in the central Congo basin. The Kulal specimens have smaller leaves and longer corolla lobes than the Congo specimens.

2. **H. zeylanica** (*Retz.*) *Thwaites* in Enum. Pl. Zeyl.: 191 (1860); Pichon in Bull. Soc. Brot. sér. 2 (27): 104 (1953); Kupicha in F.Z. 7 (2): 430 (1985); K.T.S.L.: 480, ill., map (1994); Omino in Wageningen Agric. Univ. Pap. 96 (1): 122, t. 34 (1996). Type: Sri Lanka, *Koenig* s.n. (C!, holo; BM, K-WALL!, C, MO, iso.)

Tree or shrub 1–15(–40) m high; trunk 2–30 cm in diameter; bark grey, slightly fisssured; wood very hard, dense, whitish to yellow. Leaves petiolate; blade coriaceous, elliptic, oblong to obovate, 2–21 mm long, 0.9–6.7 cm wide, apex rounded to acuminate with acumen 4–10(–17) mm long, cuneate to rounded at the base; petiole 6–17 mm long. Inflorescence terminal rarely also axillary, 10–65(–96)-flowered; bracts ovate to triangular, up to 1.3 mm long; peduncle 5–40 mm long; pedicels 3.8–10 mm long. Flowers with sepals resinous within, ovate to triangular, up to 2.5 mm long; corolla white to pale yellow, orange at throat; tube creamy green, 6–10 mm long; lobes pure white, ovate, 3.7–8.8 mm long, 1.8–3.5 mm wide, obtuse at the apex; stamens inserted at 4.5–8 mm from the base; pistil 5.5–9.5 mm long; ovary abruptly narrowing into the style, ovules 2 in each carpel; style 3.5–7.2 mm long. Fruits obovoid to globose, smooth, 13–30 mm long; 10–15 mm wide, apex rounded and sometimes stipitate at the base. Fig. 12 (p. 37).

KENYA. Northern Frontier District: Tana R. at Chewele [Chebele], 26 Oct. 1945, *Adamson* 168 (*Bally* 6068)!; Meru District: Meru National Park, Masanduku camp, *Hamilton* 93!; Kilifi District: near Hadu [Adu], Jan. 1937, *Dale* 1077!
TANZANIA. Tanga District: Mamboni beach, near Tanga, 13 Nov. 1953, *Faulkner* 1280!; Uzaramo District: Kawe, 11 km N of Dar, 15 Jan. 1971, *Harris & Jones* 5551!; Bagamoyo District: Bagamoyo, Feb. 1889, *Sacleux* 678!; Pemba, 26 Oct. 1926, *Vaughan* 885
DISTR. **K** 1, 4, 7; **T** 3, 6; **Z**, **P**; Somalia, Mozambique; India, Sri Lanka, Burma, Thailand, Cambodia, Laos, Vietnam, S China (Hainan), Malaysia and Indonesia
HAB. Coral rag forest, riverine forest, thick coastal bushland; 0–350 m
USES. Wood used for implements

SYN. *Cameraria zeylanica* Retz in Obs. Bot. 4: 24 (1786); A.DC., Prodr. 8: 389 (1844)
   Hunteria africana K. Schum in P.O.A. C: 317 (1895); T.T.C.L.: 50 (1949). Types: Tanzania, Pangani, *Stuhlmann* 77 (B†, syn.); Bagamoyo, *Stuhlmann* 206 (B†, syn.); Tanga district, Sawa, *Faulkner* 1626 (K!, neo., B, BR, FT, P, isoneo., chosen by Omino)
   H. zeylanica (Retz.) Thwaites var. *africana* (K. Schum.) Pichon in Bull. Soc. Brot. sér. 2, 27: 104 (1953); K.T.S. 47 (1961)

10. PLEIOCARPA

Benth. in G. P. 2: 699 (1876); Omino in Wageningen Agric. Univ. Pap. 96 (1): 134–157 (1996)

Shrubs or small trees, rarely lianas, with all parts glabrous except inside the corolla tube; white latex present in all parts. Leaves opposite or in whorls of 3–5. Inflorescence predominantly axillary, sometimes terminal at the same time, fasciculate, ramiflorous, 1–40-flowered, sessile; bracts very small, scale-like; pedicels 1–2 mm long. Flowers very fragrant. Sepals without colleters. Corolla pure white to yellow; tube greenish in bud; lobes overlapping to the left in bud, always shorter than the tube; ovary of 2–5 separate carpels, united at the extreme base by a disk-like thickening; pistil head ellipsoid or oblong. Fruit yellow to bright orange, smooth to rugose, of 2–5 separate mericarps (2 in our area), with a rounded to hooked apex. Seeds brown, sometimes angular; embryo straight, spatulate.

Leaves opposite; inflorescence 10–15-flowered; corolla lobes
5–10.5 mm long ·   1. *P. bicarpellata*
Leaves opposite or in whorls of 3–5; inflorescence 10–40-
flowered; corolla lobes 1.3–4.8 mm long · · · · · · · · · · · · · · ·   2. *P. pycnantha*

1. **P. bicarpellata** *Stapf* in K.B. 1894: 21 (1894) & in F.T.A. 4 (1): 99 (1902); Pichon in Bull. Soc. Brot. sér. 2 (27): 125, t. 3/8–10 (1953); Huber in F.W.T.A. ed. 2, 2: 63 (1963); Omino in Wageningen Agric. Univ. Pap. 96 (1): 135, t. 36 (1996). Lectotype: Cameroon, Mt Cameroon, *Mann* 1213 (K!, lecto.; P!, iso., chosen by Omino)

Shrub or small tree 1.5–8 m high; trunk very slender, smooth; bark and branches smooth. Leaves opposite, petiolate; blade elliptic to oblong, 5–17 cm long, 1.8–7.2 cm wide, apex acuminate with acumen 10–22 mm long, acute or cuneate at the base; petiole 4–11(–14) mm long. Inflorescence axillary, sometimes terminal at the same time, 1.5–2.5 cm long, 10(–15)-flowered. Flowers very fragrant; sepals ovate, up to 2.8 mm long; corolla pure white; tube (9.2–)10–15 mm long; lobes narrowly oblong to broadly ovate, 5–10.5 mm long, 1.5–5.5 mm wide; stamens inserted at 7.5–12 mm from the base; pistil 7.5–12.5 mm long; style 5.5–11 mm long; pistil-head oblong or ellipsoid; ovules 2 in each carpel. Fruits of 2 separate mericarps, these obovoid or ellipsoid, 11–18 mm long, 7–12 mm wide, rounded at the apex. Seeds ellipsoid, oblong or subglobose, 9–11 mm long, 3–7 mm in diameter.

KENYA. Teita District: Chawia Forest, *Omino* 82! & *Omino* 157! & Kasigau, Bungule route, 18 Nov. 1994, *Luke* 4202!
DISTR. **K** 7; Cameroon, Gabon, Congo (Kinshasa), Angola
HAB. Moist forest; 1500–1800 m
USES. None recorded

2. **P. pycnantha** (*K. Schum.*) *Stapf* in F.T.A. 4, 1: 99 (1902); I.T.U. ed. 2: 29 (1952); Pichon in Bull. Soc. Brot. sér. 2 (27): 128, t. 4/1 (1953); Huber in F.W.T.A. ed. 2, 2: 63 (1963); Hamilton, Field Guide Uganda For. Trees: 164 (1981); Kupicha in F.Z. 7 (2): 432, t. 99 (1985); K.T.S.L.: 482, ill. (1994); Omino in Wageningen Agric. Univ. Pap. 96 (1): 146 (1996). Type: Uganda, Masaka District, Sese Is., *Stuhlmann* 1216 (B†, holo., K!, fragm., iso.)

Tree 1.5–20(–30) m high; trunk 2–50 cm in diameter; bark smooth or slightly rough, pale to dark grey; wood hard, durable. Leaves opposite or in whorls of 3–5, petiolate; blade narrowly elliptic to oblong, rarely obovate, 4–23 cm long, 1.1–8 cm wide, apex acute to acuminate with acumen 4–12 mm long, cuneate at the base or decurrent into

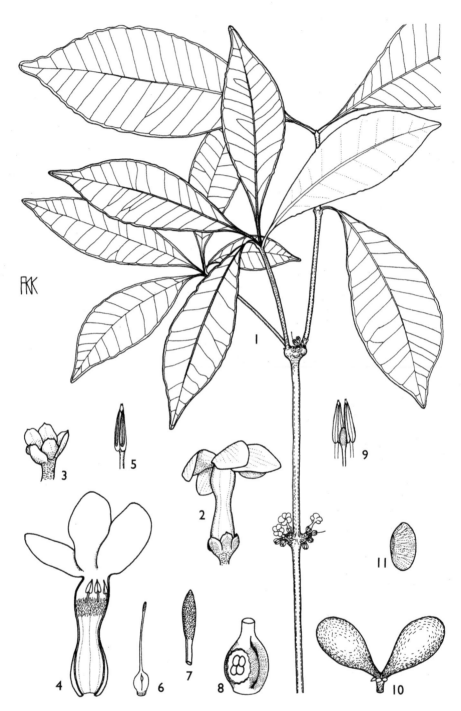

FIG. 13. *PLEIOCARPA PYCNANTHA* — **1**, habit, × ²⁄₃; **2**, flower, × ²⁄₃; **3**, calyx, × 3; **4**, part of corolla opened out, × 4; **5**, stamen, × 12; **6**, gynoecium, × 4; **7**, enlargement of 6 showing papillose clavuncle, × 24; **8**, ovary with part of wall removed (diagrammatic); **9**, relative positions of anthers and clavuncle at anthesis, × 12; **10**, fruit, × 1; **11**, seed × 1. 1–8 from *Gomes e Sousa* 4407; 9 from *Goldsmith* 38/65; 10–11 from *Goldsmith* 144/67. From F.Z., drawn by F.K. Kupicha.

the petiole; petiole 5–20 mm long. Inflorescence axillary, rarely terminal, 10–30(–40)-flowered; pedicels 1–3 mm long. Flowers white, fragrant or not; sepals broadly ovate to elliptic, up to 2.5 mm long; corolla white to orange-yellow; tube greenish white, 6.2–10.5 mm long; lobes ovate or elliptic to suborbicular, 1.3–4.5(–4.8) mm long, up to 3.5 mm wide; stamens inserted 4.6–8(–9.5) mm from the base; pistil glabrous, 4.6–8.6(–9.7) mm long; pistil head ellipsoid to ovoid. Fruits yellow, subglobose to pyriform, sometimes stipitate, 13–23(–30) mm long, up to 24 mm in diameter; seeds ellipsoid to oblong, 6.5–13.5 × 4.5–9.5 × 2–4 mm. Fig. 13 (p. 40).

UGANDA. Toro District: Kibale Forest Reserve, Mar. 1943, *Blair-Simpson* 5245!; Ankole District: Kalinzu Forest, 27 Oct. 1969, *Synnott* 410!; Mengo District: near Budo, Mar. 1932, *Eggeling* 517!
KENYA. Meru District: Meru National Park, Kindani camp site, 28 Sept 1980, *Hamilton* 738!; Kwale District: Mwele Mdogo Forest, Shimba Hills, 6 Jan. 1988, *Luke* 893A!; Lamu District: Witu Forest Reserve, 18 Nov. 1988, *Luke & Robertson* 1405!
TANZANIA. Bukoba District: Kaigi, 1935, *Gillman* 352!; Mpanda District: Kungwe Mt, 11 Sept. 1959, *Harley* 9587A!; Lindi District: Mchinjiri, Rondo plateau, Mar. 1952, *Semsei* 679!; Zanzibar, Chwaka mile 17, Mar. 1962, *Faulkner* 3022
DISTR. U 2, 4; K 4, 7; T 1, 3, 4, 6, 8; Z; tropical Africa from Senegal in the W to Angola, Zambia and Mozambique
HAB. Evergreen forest, riverine or swamp forest; 0–2300 m
USES. Wood hard, used for construction and implements; roots used in palm wine

SYN. *Hunteria pycnantha* K. Schum. in E.J. 23: 222 (1896)
  Pleiocarpa tubicina Stapf in K.B. 1898: 304 (1898); De Wild. & Durand in Ann. Mus. Congo Bot., ser. 2, 1, 1: 37 (1899) & 1, 2: 150 (1901). Type: Congo (Kinshasa), lower Congo R., *Dewèvre* 945 (BR, holo., K!, iso.)
  P. microcarpa Stapf in F.T.A. 4, 1: 102 (1902). Type: Congo (Kinshasa), Niamniam land, Mbrowole R., *Schweinfurth* 3073 (K!, holo.)
  P. bagshawei S. Moore in J. Bot. 45: 49 (1907). Type: Uganda, Toro Distr., Dura [Durro] Forest, *Bagshawe* 1086 (BM!, holo.)
  P. pycnantha (K. Schum.) Stapf var. *tubicina* (Stapf) Pichon in Bull. Soc. Brot. sér. 2 (27): 132, map c (1953); Huber in F.W.T.A. ed. 2, 2: 63 (1963)
  P. sp. near *swynnertonii* of T.T.C.L.: 54 (1949)

11. VOACANGA

Thouars, Gen. Nov. Madag.10 (1806); Leeuwenberg in Wageningen Agric. Univ. Pap. 85, 3: 9–122 (1985)

Shrubs or trees, repeatedly dichotomously branched; bark and branchlets with some white latex. Leaves opposite; petiole or leaf bases of a pair connate into a short ochrea, with a single row of colleters in the axils. Inflorescences cymose, in pairs at the forks of branches; bracts deciduous, rarely persistent, leaving conspicuous scars. Flowers often fragrant. Calyx green, usually shed with the corolla, united for over ²/₃, inside with colleters. Corolla creamy or yellow; tube shorter or only slightly longer than than the calyx, twisted; lobes in bud overlapping to the left; stamens exserted or included; anthers introrse, sessile, narrowly triangular, acuminate at the sterile apex, sagittate at the base, glabrous; ovary broadly ovoid, of two carpels, free or connate at the base, rarely entirely fused; disk present; pistil head with a ring at the base, obovoid, coherent with the connectives of the anthers; stigmoid apex short; style and stigma shed with the corolla. Fruits of two fleshy carpels, free, less often partly of wholly united. Seeds surrounded by a yellow or orange pulpy aril, numerous; endosperm copious, starchy, creamy to white, ruminate, surrounding the spatulate creamy white embryo.

An Old World genus of 12 species, 7 in Africa and 5 in SE Asia.

1. Leaves usually sessile, elliptic, bluntly acuminate at the apex;
    corolla tube 7–15 mm long; lobes obovate or elliptic, 7–16
    mm wide, rounded at the apex · · · · · · · · · · · · · · · · ·   1. *V. africana*
    Leaves petiolate, narrowly obovate, obtuse or rounded at the
    apex; corolla tube 17–23 mm long; lobes broadly obcordate,
    28–43 mm wide, emarginate at the apex · · · · · · · · · · · ·   2. *V. thouarsii*

1. **V. africana** Stapf in J.L.S. 30: 87 (1894) & in F.T.A. 4, 1: 157 (1902); I.T.U. ed.
2: 30 (1952); Huber in F.W.T.A. ed. 2, 2: 67 (1963); Leeuwenberg in F.Z. 7 (2): 435,
t. 100 (1985) & in Wageningen Agric. Univ. Pap. 85, 3: 12, t. 1, photo. 1 (1985);
K.T.S.L.: 487 (1994). Type: Sierra Leone, Bafodeya, *Scott Elliot* 5484 (BM, lecto., K,
iso., chosen by Leeuwenberg)

Shrub-like tree or shrub, 1–10(–25) m high; trunk terete, 2–30(–40) cm in
diameter; bark pale grey-brown, smooth or shallowly fissured; branches lenticellate,
branchlets glabrous, puberulous, or pubescent. Leaves petiolate; blade elliptic or
narrowly elliptic, 7–42 cm long, 3–20 cm wide, bluntly acuminate, rarely acute or
obtuse at the apex, cuneate or decurrent into the petiole at the base, glabrous,
sometimes pubescent beneath and on midrib above; petiole 0–2 cm long, glabrous
to pubescent, ochrea not widened into intrapetiolar stipules. Inflorescence 6–25 cm
long, many-flowered, glabrous to sparsely pubescent in all parts; peduncle 5–18 cm
long; bracts deciduous, ovate, obtuse; pedicels 3–20 mm long. Flowers malodorous;
sepals 7–19 mm long, deciduous before the fruit develops, glabrous or puberulous;
corolla creamy, greenish creamy, yellow or less often white; tube 7–15 mm long,
almost cylindrical, twisted from 2–3 mm above the base; lobes twisted in bud,
(narrowly) obovate or elliptic, 12–37 mm long, 7–16 mm wide, rounded or obtuse
at the apex, entire, spreading, often recurved later; stamens exserted for up to 1.2
mm, rarely just included; anthers glabrous, usually twisted with corolla; pistil 7–12.5
mm long; ovules about 200 in each carpel; disk entire, ring-shaped, 0.8–1.2 mm
high; style split at the base, 4–8 mm long. Fruit mericarps separate, only one
developed, dark and blotched with very pale green, obliquely subglobose, often
wider than long, laterally compressed 3–8 cm long, 3–8 cm wide; pericarp 5–15 mm
thick, cream and orange.

UGANDA. West Nile District: Laropi, W Madi, *Eggeling* 1792; Mengo District: Entebbe, *Mahon* 1
KENYA. Kwale District: Gazi–Shimoni road, on bridge over Ramisi R., 20 Aug. 1953, *Drummond
& Hemsley* 3902! & near Makweyeni, 19 Aug. 1993, *Luke* 3829! & Vanga, *Graham* 2210
TANZANIA. Lushoto District: Mombo Forest Reserve, 21 Dec. 1956, *Muge* 15; Kilosa District:
Kilosa, 1.5 km from Kondoa R. bridge, 20 Sept. 1969, *Mwasumbi & Mwakalasi* 936!; Rungwe
District: Unyakyusa area, 6 Dec. 1974, *Leedal* 2376!
DISTR. **U** 1, 4; **K** 7; **T** 3, 4, 6–8; tropical Africa from Senegal to Central African Republic to
Congo (Kinshasa), Burundi and Angola, to Zambia, Malawi and Mozambique
HAB. Forest, riverine forest; 0–1200 m
USES. Major medicinal for geriatric patients; wood for building poles

SYN. *V. schweinfurthii* Stapf in K.B. 1894: 21 (1894) & in F.T.A. 4, 1: 155 (1902). Type: Congo
(Kinshasa), Niamniam land, Dungu, Turu R., *Schweinfurth* 3326 (B†, holo., K, lecto., K,
P, iso., chosen by Leeuwenberg)
V. angustifolia K. Schum. in P.O.A. C: 317 (1895). Type: Tanzania, Tabora District, Kavenda
R., *Böhm* 60a (B†, holo.)
V. boehmii K. Schum. in P.O.A. C: 317 (1895); T.T.C.L.: 57 (1949). Type: Tanzania, Tabora
District, Kavenda [Kavinda] R., Isimbira, *Böhm* 37a (B†, holo., K, lecto., K, P, iso.)
V. dichotoma K. Schum. in P.O.A. C: 317 (1895); T.T.C.L.: 57 (1949). Type: Tanzania, Moshi
District, Kiboshowald, *Volkens* 1618 (B†, syn.) & Marangu, *Volkens* 2076 (B†, syn.)
V. lutescens Stapf in F.T.A. 4, 1: 157 (1902); T.T.C.L.: 57 (1949); K.T.S.: 49 (1961). Lectotype:
Mozambique, Manica e Sofala, between Lupata & Sena, *Kirk* 31 (K, lecto.), chosen by
Leeuwenberg)
V. bequaertii De Wild., Pl. Bequaert. 1: 402 (1922). Type: Congo (Kinshasa), between
Walikale & Lubutu, *Bequaert* 6577 (BR, holo.)

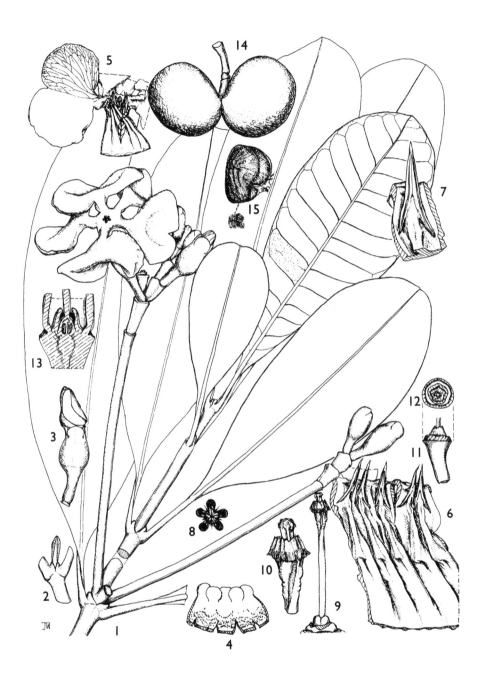

Fig. 14. *VOACANGA THOUARSII* — **1**, flowering branch, × ²/₃; **2**, apex of branchlet, × ²/₃; **3**, flower bud, × ²/₃; **4**, calyx inside, × ²/₃; **5–6**, opened corollas, × ²/₃; **7**, stamen, × 4; **8**, corolla mouth showing anther apices × 2; **9**, pistil, × 2; **10**, pistil head, × 4; **11**, pistil base with high disk, × 1; **12**; pistil base above view, × 1; **13**, section of flower base without calyx, × 2; **14**, fruit, × ¹/₃; **15**, seed, × 2. All from *Leeuwenberg*, 1; 5 from 11904; 2 from 11206; 3–4 & 6–13 from 9371; 14–15 from 10726. From Meded. Landbouwhogeschool 85(3), drawn by J. Williamson, reproduced by permission.

V. africana Stapf var. *lutescens* (Stapf) Pichon in Bull. Mus. Nat. Hist. Nat. Paris ser. 2, 19: 412 (1947)

V. africana Stapf var. *auriculata* Pichon in Bull. Mus. Nat. Hist. Nat. Paris ser. 2, 19: 412 (1947). Type: Tanzania, Sagara (?= Segera, Pangani District?), *Sacleux* 2222 (P, holo.)

V. densiflora Engl., nom. nud.; T.T.C.L.: 57 (1949)

2. **V. thouarsii** *Roem. & Schult.*, Syst. Veg. 4: 439 (1819); Stapf in F.T.A. 4, 1: 154 (1902); T.T.C.L.: 47 (1949); Codd in F.S.A. 26: 273 (1963); Huber in F.W.T.A. ed. 2, 2: 67 (1963); Hamilton, Field Guide Uganda For. Trees: 166, ill. (1981); Leeuwenberg in F.Z. 7 (2): 436 (1985) & in Wageningen Agric. Univ. Pap. 85, 3: 39, t. 7 (1985); K.T.S.L.: 488, ill. (1994). Type: Madagascar, sine loc., Herb. *Du Petit-Thouars* s.n. (P, holo.)

Tree 2–15(–20) m high; trunk terete, 4–40(–80) cm in diameter; bark pale grey brown, smooth, with small lenticels, and some white latex; branchlets glabrous or minutely puberulous, with much latex. Leaves petiolate; blade narrowly obovate, 6–25 cm long, 2–9 cm wide, obtuse or rounded at the apex, attenuate at the base, glabrous, with numerous pits on both surfaces; petiole 8–25 mm long, glabrous or minutely puberulous at the base, ochrea widened into intrapetiolar stipules. Inflorescence up to 21 cm long corymbs, few-flowered; peduncle 5–20 cm long, stout, glabrous; bracts deciduous, ovate, apex rounded, up to 10 mm long; pedicels 8–15 mm long, minutely puberulous above. Flowers sweetly scented; sepals connate for over half their length, fleshy, broadly ovate, rounded; tube often pale green, creamy or white, fleshy, almost cylindrical, 17–23 mm long, twisted; lobes pale yellow, broadly obcordate, not twisted, 19–30 mm long, 28–43 mm wide, narrowed at the base, emarginate at the apex; stamens barely exserted or included 3–4 mm below the mouth of the corolla; disk annular, 2–4.5 mm high, shallowly to deeply 5-lobed; pistil 17–23 mm long; ovary of free carpels, each with ± 80 ovules; style 12–19 mm long. Fruits spotted pale and dark green, of two free subglobose mericarps, 4–10 cm in diameter; pericarp creamy in section, 5–15 mm thick; aril orange; seeds dark brown, obliquely ovoid or ellipsoid, 8–9.5 mm long, densely papillose and shallowly grooved all over. Fig. 14 (p. 43).

UGANDA. Ankole District: Buhweju, Katera R., 23 July 1969, *Synnott* 374!; Mengo District: Nambigirwa swamp, Entebbe, Jan. 1932, *Eggeling* 151!; Masaka District: Bale, Lake Nabugabo, 5 May 1969, *Lye* 2774!

KENYA. North Kavirondo: Malava Forest, 27 Nov. 1969, *Faden & Evans* 69/2061; Central Kavirondo: Soso, Kalanyo, Yala, 16 Sept. 1972, *Kokwaro* 3155 & Gem, Yala, 2 July 1972, *Odera* 15!

TANZANIA. Bukoba District: Nyakato, April 1935, *Gillman* 257!; Lushoto District: Monga, 15 Feb. 1917, *Zimmermann* 7750!; Songea District: ± 6.5 km E of Gumbiro, 25 Jan 1956, *Milne-Redhead & Taylor* 8427!

DISTR. **U** 2, 4; **K** 5; **T** 1, 3, 6–8; Tropical Africa from Senegal to Sudan and Angola and Malawi to S Africa, Madagascar

HAB. Riverine forest or bush, swamps (where it may be dominant); 0–1600 m

USES. Major pharmaceutical medicinal for geriatric patients

SYN. *V. obtusa* K. Schum. in E. & P. Pf. 4 (2): 149 (1895); Stapf in F.T.A. 4, 1: 153 (1902); T.T.C.L.: 57 (1949); I.T.U. ed. 2: 31 (1952); K.T.S.: 49 (1961). Type: Congo (Kinshasa), Niamniam land, Mbruole R., *Schweinfurth* 3741 (B†, holo., K, lecto., P, S, iso., chosen by Leeuwenberg)

V. thouarsii Roem. & Schult. var. *obtusa* (K. Schum.) Pichon in Bull. Mus. Nat. Hist. Nat. Paris ser. 2, 19: 412 (1947)

12. TABERNAEMONTANA

L., Sp. Pl.: 210 (1753); Leeuwenberg in Revision Tabernaemontana 1: 1–204 (1991)

Conopharyngia G. Don, Gen. Syst. 4: 94 (1837)

Shrubs or trees repeatedly dichotomously branched; latex present in most parts; bark smooth or rough, usually with large lenticels; branches with large lenticels and conspicuous leaf scars; branchlets terete, glabrous. Leaves opposite; petiole glabrous, those of a pair usually connate into a conspicuous ochrea, these frequently widened into intrapetiolar stipules, with colleters in axils; blade glabrous on both sides (occasionally pubescent beneath in *T. elegans*), entire. Inflorescence terminal or in forks, corymbose; bracts deciduous, with colleters in axils, leaving large scars. Flowers 5-merous, actinomorphic, sweet-scented. Sepals imbricate in bud. Corolla tube thin or thick and fleshy, twisted or not; lobes overlapping to the left, twisted, folded inwards, obliquely ellipsoid and falcate, curved to the right, often auriculate at the left side at the base; stamens included; filaments short or reduced to ridges; anthers sessile, narrowly triangular, acuminate at the sterile apex, sagittate at the base, with connectives adnate to the corolla; ovary superior, of two carpels, barely to distinctly connate at the base, connate at the apex by the style; pistil head not coherent with anthers, subglobose or cylindrical with 5 suborbicular or elliptic lobes at the apex and an entire undulate or lobed ring at the base. Fruit of two separate or rarely basally united mericarps, subglobose to pod-like, baccate, dehiscent or not; aril pulpy, white, orange or red. Seeds brown or black, with a deep groove on the hilar sides; endosperm copious, starchy, white, ruminate; embryo spatulate .

A genus of about 110 species, circumtropical.

The following species have been introduced:
T. divaricata (L.) Roem. & Schult., syn. *T. coronaria* (Jacq.) Willd.; U.O.P.Z.: 458, plate (1949); a form with double flowers has been cultivated in Zanzibar.
T. pandacaqui Lam., a shrub or small tree, has been introduced from SE Asia: T 6, Ras Kutani, Feb. 1995, *Luke & Luke* 4284

1. Corolla tube 5–7 mm long, up to 2.4 mm wide; fruit with pale brown warts, apiculate ·   1. *T. elegans*
   Corolla tube > 10 mm long, at least 5 mm wide; fruit smooth or dotted, rounded (or acuminate in *T. odoratissima*) · · · · · · · · · · · · · · · 2
2. Corolla tube > 45 mm long, lobes narrowly oblong; stamens with apex below half the length of corolla tube; fruit pod-like or subglobose ·   2. *T. odoratissima*
   Corolla tube < 45 mm long; lobes obliquely elliptic; stamens with apex above half the length of corolla tube; fruit obliquely ellipsoid or subglobose · 3
3. Stamens inserted 6–8 mm above the corolla base; pistil 7–10 mm long; fruit obliquely ellipsoid, ridged · · · · · · · · · ·   5. *T. ventricosa*
   Stamens inserted 8–15 mm above the corolla base; pistil 13–20 mm long; fruit subglobose, dotted · 4
4. Mature bud with a broadly ovoid head with a blunt apex, comparatively smaller than tube; corolla tube angular, barely twisted at the base ·   3. *T. pachysiphon*
   Mature bud with an ovoid head with a subacute apex, conspicuously wider than the tube; corolla tube not angular, twisted a whole 360 turn over its length · · · · · ·   4. *T. stapfiana*

1. **T. elegans** *Stapf* in K.B. 1894: 24 (1894); Codd in F.S.A. 26: 270, t. 391 (1963); Leeuwenberg in F.Z. 7 (2): 439 (1985) & Revision Tabernaemontana 1: 32, t. 7, photo 1–2 (1991); K.T.S.L.: 486, map (1994). Type: Mozambique, Maputo, Delagoa Bay, *Monteiro* 55 (K!, lecto., G, P, isolecto., chosen by Leeuwenberg)

Small tree or shrub 1.5–15 m high, trunk 5–30 cm in diameter; bark pale brown, corky, fissured. Leaves petiolate; blade elliptic or narrowly elliptic, 6–23 cm long, 2–8 cm wide, apex acute, obtuse or acuminate, cuneate at the base or decurrent into the petiole, occasionally pubescent beneath; petiole 7–30 mm long. Inflorescence 5–20 cm long, many-flowered, lax, glabrous or occasionally with scattered short hairs in all parts; peduncle 1–8.5 cm long; pedicels 2–6 mm long. Flowers with sepals suborbicular to broadly ovate, 1.2–2.5 mm long, 1.2–2.5 mm wide, rounded; corolla creamy or pale yellow, white distally; tube almost cylindrical, not twisted, 5–7 mm long; lobes spreading, obliquely elliptic, slightly falcate, 8–15 mm long, 3–7 mm wide, rounded, subauriculate at the left side of the base, entire; stamens with apex 1–2 mm below the mouth of the corolla tube, inserted 2–2.7 mm above the corolla base; pistil 3.5–4.2 mm long, glabrous; ovary up to 1.6 mm long; ovules 35–60 in each carpel. Fruit green, with conspicuous pale brown warts, obliquely ovoid or ellipsoid, 5–8 cm long, 4–6.5 cm in diameter, apiculate, 3–ridged; pericarp 5–15 mm thick; aril orange; seeds dark brown, coffee-bean-like, 14–15 mm long, reticulately grooved, papillose.

KENYA. Kwale District: Shimba Hills, Marere pumping station, 3 Apr. 1968, *Magogo & Glover* 746!; Kilifi District: Mazeras, Apr. 1930, *Graham* 2309!; Lamu District: SW edge of Boni Forest, 62 km ESE of Mararani, *Kuchar* 13483!
TANZANIA. Uzaramo District: Mzizima farm, Dar es Salaam, 13 July 1971, *Hansen* 360!; Bagamoyo District: Bana Forest Reserve, 3 Aug. 1968, *Shabani* 80!; Kilwa District: 5 km NNW of Kingupira, 21 Nov. 1975, *Vollesen* 3026!
DISTR. **K** 7; **T** 3, 6, 8; Somalia, Malawi, Mozambique, Zimbabwe, Swaziland and South Africa
HAB. Moist or dry forest, coastal bushland, evergreen thicket; 0–850 m
USES. Cultivated in several botanic gardens

SYN. *Conopharyngia elegans* (Stapf) Stapf in F.T.A. 4 (1): 149 (1902); T.T.C.L.: 49 (1949); K.T.S.: 45 (1961)
     Leptopharyngia elegans (Stapf) Boiteau in Adansonia II (16): 276 (1976)

2. **T. odoratissima** (*Stapf*) *Leeuwenberg* in Pelletier, Alkaloids 1: 338 (1983) & in Revision Tabernaemontana 1: 49, t. 12 (1991); Hamilton, Field Guide Uganda For. Trees: 164 (1981), *comb. non rite publ.* Type: Uganda, W Ankole, *Dawe* 352 (K, holo.)

Tree 5–15 m high; branches pale to dark brown; branchlets glabrous. Leaves petiolate; blade elliptic or narrowly elliptic, 9–27 cm long, 2.5–10.5 cm wide, apiculate to obtuse at the apex, cuneate at the base, often with scattered black dots beneath, with 7–15 pairs of rather straight secondary veins; petiole 5–16 mm long. Inflorescence 10–20 cm long, few- to many-flowered, rather congested, glabrous in all parts; peduncle robust, 4–7 cm long; pedicels 5–24 mm long. Flowers opening during the night; sepals ovate, 4–7 mm long, 3–4 mm wide, rounded, ciliate; corolla white, fleshy, forming a small depressed globose or ovoid head in mature bud, with a blunt or rounded apex; tube almost cylindrical, 45–100 mm long, twisted; lobes spreading, narrowly oblong, more or less falcate, 20–55 mm long, 5–10 mm wide, acute or obtuse, auriculate at the left side of the base; stamens with apex 29–83 mm below mouth of corolla tube, inserted 1.5–5 mm from the base; pistil 5.5–5.8 mm long, glabrous; style very short, 0.7–2.5 mm long; ovary ovoid or subglobose; ovules 100–130 in each carpel. Fruit mericarps subglobose and 50 mm long, 45 mm in diameter or pod-like and 90 mm long, 15 mm wide, acuminate and recurved; pericarp about 4–10 mm thick; seeds dark brown, obliquely ovoid, 9–11 mm long, longitidinally grooved, dull, densely papillose.

Uganda. Ankole District: Bunyaruguru, Kalinzu Forest, 7 Jun. 1969, *Synnott* 335!; Toro District: Katojo, Kibale Forest, 28 Aug. 1970, *Katende* 513; Mengo District: Mpanga Forest Reserve, 1906, *Bagshawe* 999
Tanzania. Ulanga District: 3 km from Mahenge, 6 Dec. 1958, *Haerdi* s.n.! & between Kidatu and Sonjo, *Harris & Pocs* 4281; Morogoro District: Kimboza Forest Reserve, *Padwa* 333
Distr. U 2, 4; T 6, 7; Congo (Kinshasa) and Rwanda
Hab. Moist forest, woodland on rock; 450–1850 m
Uses. None recorded

Syn. *Gabunia odoratissima* Stapf in J.L.S. 37: 526 (1906); I.T.U. ed. 2: 28 (1952)

3. **T. pachysiphon** *Stapf* in K.B. 1894: 22 (1894); Huber in F.W.T.A. ed. 2, 2: 66 (1963); Leeuwenberg in Journ. Ethnopharm. 10: 16 (1984) & in F.Z. 7 (2): 440, t. 101 (1985) & in Revision Tabernaemontana 1: 51, t. 13 (1991); K.T.S.L.: 486 (1994). Type: Nigeria, Lower Niger, Onitsha, *Barter* 1320 (K, holo.)

Shrub or small tree 2–15 m high; trunk 4–40 cm in diameter; bark pale brown or grey brown, with large lenticels. Leaves petiolate; blade broadly to narrowly elliptic or obovate, 10–42(–50) cm long, 5–20(–26) cm wide, acuminate to acute at the apex, cuneate at the base; petiole 6–25 mm long, glabrous. Inflorescence 8–26 cm long, few to many-flowered, glabrous in all parts; peduncle glabrous, 3–14 cm long; pedicels 8–22 mm long. Flowers with sepals suborbicular or ovate, 4–7 mm long and wide, ciliate, with colleters within; corolla thick, fleshy; mature bud with a comparatively small broadly ovoid head with a blunt apex; tube pale green, almost cylindrical, throat yellow, 18–35(–42) mm long, 5–angular, sometimes slightly twisted at the base; lobes spreading, later recurved, white, sometimes pale yellow, obliquely elliptic, more or less falcate, 14–50 mm long, 6–18(–27) mm wide, rounded; stamens with apex up to 12 mm below mouth of corolla tube, inserted 8–14 mm above the corolla base; pistil glabrous, 15–20 mm long; ovary almost cylindrical; ovules about 100 in each carpel; style 7–10 mm long. Fruits pale green, often dotted, obliquely subglobose, 7–15 cm long, 6–18 cm in diameter, rounded, several- to many-seeded; pericarp 2.3–4 cm thick; aril white; seeds dark brown, coffee-bean-like, obliquely ellipsoid, 11–14 mm long. Fig. 15 (p. 48).

Uganda. Kigezi District: Amahenge, Kinkizi, March 1946, *Purseglove* 2021!; Mengo District: Lakeshore forest, Entebbe, *Eggeling* 211!; Masaka District: Kalangala, Bugala Is., 4 Mar. 1951, *Phillips* 432!
Kenya. Kwale District: Shimba Hills, Longomwagandi forest, 28 Feb. 1968, *Magogo & Glover* 201! & Buda Forest, W Msambweni, 13 Aug. 1976, *Saufferer* 748! & Mwele Mdogo Forest, Dec., *Moomaw* 1059!
Tanzania. Moshi District: Kilimanjaro, 22 Nov. 1901, *Uhlig* 644!; Lushoto District: Amani, 27 Jun. 1970, *Hansen* 121!; Newala District: Chilangala, Makandi–Plateau, 12 Nov. 1958, *Haerdi* 5!; Zanzibar, Jozani, July 1972, *Robbins* 47
Distr. U 2, 4; **K** 7; **T** 1–4, 6, 8; **Z, P**; widespread in tropical Africa from Ghana to Sudan and Angola to Malawi
Hab. Moist forest and forest margins, riverine and gallery forest; 0–2000 m
Uses. Fruits eaten by chimpanzees

Syn. *T. holstii* K. Schum. in P.O.A. C: 317 (1895); Hamilton, Field Guide Uganda For. Trees: 164 (1981). Lectotype: Tanzania, Lushoto District, Derema, *Holst* 2247 (K, lecto., chosen by Leeuwenberg)
    Voacanga dichotoma K. Schum. in P.O.A. C: 317 (1895). Type: Tanzania, Moshi District, Marangu, *Volkens* 2076 (HBG, lecto., BR, isolecto., chosen by Leeuwenberg)
    Conopharyngia holstii (K. Schum.) Stapf in F.T.A. 4, 1: 146 (1902); T.T.C.L.: 49 (1949); I.T.U. ed. 2: 26 (1952); K.T.S.: 46 (1961)
    C. pachysiphon (Stapf) Stapf in F.T.A. 4, 1: 146 (1902)
    C. angolensis (Stapf) Stapf in F.T.A. 4, 1: 146 (1902); T.T.C.L.: 49 (1949). Type: Angola, Pungo Andongo, *Welwitsch* 5989 (BM, holo, G, K, LISU, P, iso.)

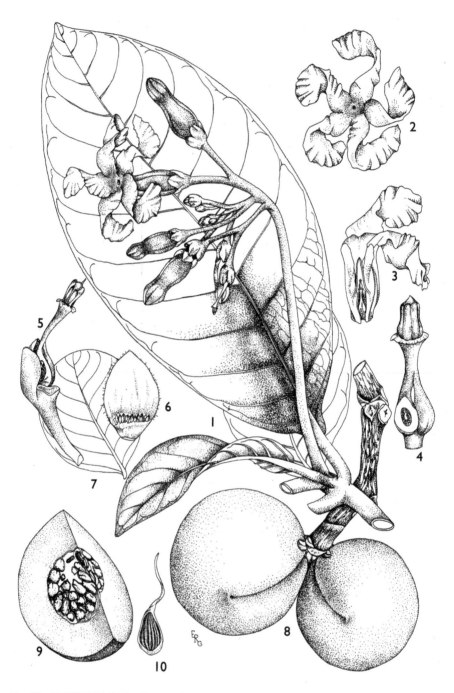

Fig. 15. *TABERNAEMONTANA PACHYSIPHON* — **1**, habit, × ²/₃; **2**, flower from above, × ¹/₂; **3**, opened corolla, × ¹/₂; **4**, young pistil with ovary in section, × 3; **5**, calyx with old pistil, ovary in section, × 1¹/₂; **6**, sepal, × 3; **7**, smaller leaf of a strongly unequal pair, × ¹/₄; **8**, fruit, × ¹/₄; **9**, mericarp in section, × ¹/₂; **10**, seed with aril partly removed, × 1. 1–4 from *Leeuwenberg* 11242; 5–6 from *Leeuwenberg* 11933; 7 from *Okafor & Latilo* FHI 57181; 8 from *Leeuwenberg* 11905; 9 from *Lap* 239; 10 from *Leeuwenberg* 11044. From Leeuwenberg, Rev. Tabernaemontana, drawn by E. Riemer-Gerhardt.

4. T. stapfiana *Britten* in Trans. Linn. Soc. 2 (4): 25 (1894); Leeuwenberg in Journ. Ethnopharm. 10: 18 (1984) & in F.Z. 7 (2): 442 (1985) & in Revision Tabernaemontana 1: 63, t. 16 (1991); K.T.S.L.: 486, map (1994). Type: Malawi, Mt Mulanje, *Whyte* 87 (BM, holo.)

Tree 5–25(–35) m high; trunk 25–90 cm in diameter; bark pale to dark grey brown, rough, thick, corky; wood yellow or light brown. Leaves petiolate; blade narrowly elliptic to slightly obovate, 12–42 cm long, 3–17 cm wide, acuminate, apiculate or rounded at the apex, cuneate or decurrent at the base, with scattered dots beneath; petiole 30 mm long. Inflorescence lax, 10–28 cm long, few- to many-flowered, glabrous in all parts; peduncle 3–15 cm long; pedicels 5–30 mm long. Flowers with sepals suborbicular to oblong, 5–7 mm long, ciliate or not, with colleters within; corolla white, thick, fleshy; mature bud with a comparatively large ovoid head conspicuously wider than the tube with a subacute apex; tube pale green, almost cylindrical, throat pale yellow, 21–42 mm long, not angular, twisted over the entire length of the tube; lobes spreading, later recurved, obliquely elliptic, sometimes falcate, 17–60 mm long, 8–35 mm wide, rounded, obscurely auriculate at the left side of the base; stamens included for 2–15 mm, inserted 11–15 mm from the base; pistil glabrous, 13–17 mm long; style 5–10 mm long; ovary almost cylindrical; ovules 100–200 in each carpel. Fruits dark green, densely speckled yellow or white, subglobose, obliquely ovoid and angular when young, 10–20 cm long, 8–20 cm in diameter, rounded, dehiscent, several–many-seeded; pericarp 2.5–6 cm thick; aril orange; seeds dark brown, slightly coffee-bean-like, obliquely ellipsoid, 15–21 mm long.

UGANDA. Kigezi District: Lake Mutanda, Oct. 1940, *Eggeling* 4162! & Oct. 1947, *Purseglove* 2509!; Mbale District: Bumoni, Mt Elgon, *Snowden* 860!
KENYA. South Nyeri District: S Mt Kenya, Kamweti track, between Thiba fishing camp and crossing of Thiba R., 31 Jan. 1971, *Faden* 71/86!; North Kavirondo District: NE of Kakamega Forest Station, 28 Nov. 1971, *Higgins & Magogo* 11752! Teita District: Taita Hills, SE end of Ngangao Forest, 9 Feb. 1966, *Gillett et al.*!
TANZANIA. Moshi District: Weru Weru forest road, 1 Jan. 1967, *Bigger* 775!; Lushoto District: Jaegertal valley, 1 km NE of Lushoto, 21 Jun. 1953, *Drummond & Hemsley* 2967!; Njombe District: Mdando Forest Reserve, 15 Nov. 1966, *Gillett* 17867!
DISTR. **U** 2, 3; **K** 3–7; **T** 2, 3, 6–8; Congo (Kinshasa), Rwanda, Burundi, Malawi, Mozambique and Zimbabwe
HAB. Forest; 1400–2300 m
USES. None recorded

SYN. *Conopharyngia stapfiana* (Britten) Stapf in F.T.A. 4, 1: 147 (1902)
    C. johnstonii Stapf in F.T.A. 4, 1: 147 (1902); T.T.C.L.: 49 (1949); I.T.U. ed. 2: 27 (1952); K.T.S.: 46 (1961). Type: Kenya, Nandi Plateau, 20 Jun. 1901, *H. H. Johnston* s.n. (K!, holo.)
    C. bequaertii De Wild., Pl. Bequaert. 1: 397 (1922), as *bequaerti*. Type: Congo (Kinshasa), Ruwenzori, Lamia R. Valley, *Bequaert* 4315 (BR, holo.)
    C. johnstonii Stapf var. *grandiflora* Markgr. in N.B.G.B. 8: 496 (1923). Type: Kenya, Meru District, near Meru, *Fries* 1660 (UPS, holo, BR, K, iso.)
    Tabernaemontana johnstonii (Stapf) Pichon in Not. Syst. ed. Humbert 13: 251 (1948)
    Sarcopharyngia stapfiana (Britten) Boiteau in Bull. Mus. Hist. Nat. Paris 4 (3): 233 (1981), pro parte

5. T. ventricosa *A. DC.*, Prodr. 8: 366 (1844); Huber in F.W.T.A. ed. 2, 2: 66 (1963); Codd in F.S.A. 26: 269 (1963); Leeuwenberg in F.Z. 7 (2): 443 (1985) & in Revision Tabernaemontana 1: 71, t. 18, photo 4 (1991); K.T.S.L.: 487, map (1994). Type: South Africa, Natal, near Umgeni R., *Krauss* 146 (G, holo., BM, K, MO, TCD, iso.)

Shrub or small tree 3–15 m high; trunk 5–30 cm in diameter; bark pale brown, shallowly to deeply fissured, with large lenticels. Leaves petiolate; blade narrowly elliptic, 4–27 cm long, 1.5–10(–12) cm wide, obtusely acuminate, acute or obtuse at the apex, cuneate at the base; petiole 3–15 mm long. Inflorescence 5–23 cm long, many-flowered, glabrous in all parts; peduncle 2–15 cm long; pedicels 3–10 mm long. Flowers with sepals suborbicular or broadly ovate, 3.5–6 mm long, glabrous outside, ciliate, with a single row of colleters within; corolla white, with a pale yellow throat and greenish tube; mature bud forming a broadly ovoid head; tube almost cylindrical, 10–30 mm long, twisted just above the base; lobes spreading, later recurved, obliquely elliptic, mostly clearly falcate, 14–32 mm long, 5.5–16 mm wide, rounded, auriculate at the left side of the base, undulate; stamens barely exserted or with apex up to 14 mm below the mouth of the corolla tube, inserted 6–8 mm above the corolla base; pistil 7–10 mm long, glabrous; style 2.5–5.3 mm long; ovary almost cylindrical to broadly ovoid; ovules 70–100 in each carpel. Fruits dark green, obliquely ellipsoid, 6–7(–10) cm long, 4.5–5(–10) cm in diameter, rounded, with 2 faint lateral ridges, smooth or slightly verrucose, dehiscent, several-seeded; pericarp 7–13 mm thick; pulp white; aril orange; seed dark brown, ± coffee-bean-like, obliquely ellipsoid, 11–23 mm long.

UGANDA. Toro District: Bundibugyo, Bwamba, Oct. 1940, *Eggeling* 4057!; Kigezi District: North Maramagambo, 18 Nov. 1968, *Synnott* 209!; Masaka District: Bukoto County, 2 km E of Kayugi, 22 May 1972, *Lye* 6938!
KENYA. North Kavirondo: Kakamega forest near Ikuywa R. crossing, 6 Jan. 1968, *Perdue & Kibuwa* 9468!; Masai District: Mzima springs, Tsavo West, 17 Oct. 1966, *Gilbert* J17!; Teita District: Lume R. forest, Taveta, Sept. 1932, *Gardner* 2956!
TANZANIA. Mbulu District: W side of Lake Manyara National Park, 27 Oct. 1961, *Greenway* 10294; Morogoro District: Matibwa Forest Reserve, Dec. 1953, *Semsei* 1515!; Lindi District: Rondo Forest Reserve, 14 Feb. 1991, *Bidgood et al.* 1549!; Pemba, Ngezi Forest, Feb. 1929, *Greenway* 1470
DISTR. **U** 2, 4; **K** 3–7; **T** 1–3, 6–8; **Z**, **P**; Nigeria, Cameroon, Congo (Kinshasa), Burundi, Malawi to South Africa
HAB. Moist forest and forest margins, riverine and groundwater forest, evergreen thicket; 0–1650 m
USES. Latex used to heal wounds

SYN. *T. usambarensis* Engl. in Abh. Preuss. Akad. Wiss. 1894: 36 (1894); P.O.A. C: 316 (1895); Hamilton, Field Guide Uganda For. Trees: 166 (1981). Type: Tanzania, Lushoto District, Mashewa, Usambara, *Holst* 8810 (K, lecto., chosen by Leeuwenberg)
    Conopharyngia ventricosa (A. DC.) Stapf in F.T.A. 4, 1: 149 (1902)
    C. usambarensis (Engl.) Stapf in F.T.A. 4, 1: 148 (1902); U.O.P.Z.: 209 (1949); T.T.C.L.: 49 (1949); I.T.U. ed. 2: 27 (1952); K.T.S.: 46 (1961)
    C. rutshurensis De Wild., Pl. Bequaert. 1: 599 (1922). Type: Congo (Kinshasa), Rutshuru R., *Bequaert* 6224 (BR, holo.)
    Sarcopharyngia ventricosa (A. DC.) Boiteau in Adansonia II (16): 272 (1976)

NOTE. *Conopharyngia mborensis* Engl. in V.E. 1(1): 356 (1910), is a nomen nudum. It was said to occur in the Uluguru Mts at Mbora, as a shrub in riverine forest between 600–1000 m. It is most likely this species but no authentic material has been seen.

13. CALLICHILIA*

Stapf in F.T.A. 4(1): 130 (1902); Beentje in Meded. Landbouwhogeschool 78 (7): 1–32 (1978)

Ephippiocarpa Markgr. in Notizbl. Bot. Gart. Mus. Berl. 8 (74): 310 (1923)

Shrubs or lianas, glabrous in all parts; white latex present in most parts. Leaves opposite; petiole with a single row of colleters in the axils. Inflorescences one or two, at branch apex just below its ramification, pendulous, pedunculate, cymose, congested. Flowers actinomorphic. Corolla white; lobes in bud overlapping to the left; stamens included; anthers introrse, narrowly oblong, apiculate or obtuse,

* by H. J. Beentje

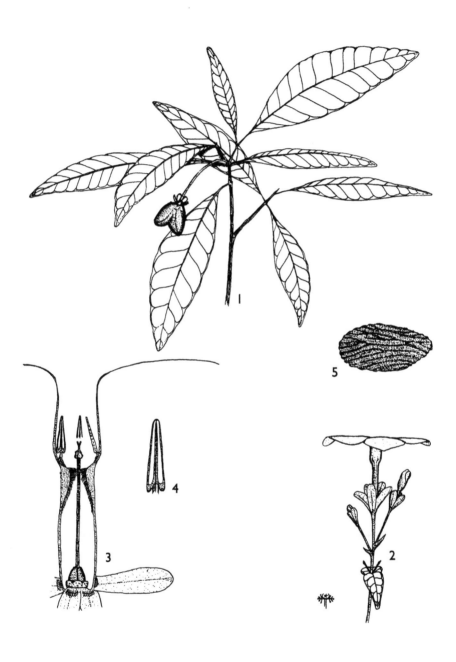

FIG. 16. *CALLICHILIA ORIENTALIS* — **1**, fruiting branch, × ¹/₂; **2**, flowering branch, × 1; **3**, opened corolla, × 3; **4**, stamen, × 6; **5**, seed, × 4. 1 from *Vahrmeyer* 434; 2–4 from *Tinley* 317; 5 from *Tinley* 213. From Meded. Landbouwhogeschool 78(7), drawn by H. Beentje.

auriculate at the base; pistil glabrous; ovary broadly ovoid, of 2 almost free carpels, surrounded at the base by a shallowly lobed almost free disk; pistil head composed of a 5-ribbed or -winged clavuncular body, each wing between the auriculae of the anther, stigma obscure. Fruit berry-like, many-seeded, the two mericarps free or partly connate, yellow to orange; wall thin, pulp fleshy; aril very thin. Seeds ovoid, with deeply pitted testa and slightly ruminate endosperm.

Seven species in tropical Africa.

C. orientalis *S. Moore* in J.L.S. 40: 139 (1911); Codd in F.S.A. 26: 272 (1963); Beentje in Meded. Landbouwhogeschool 78 (7): 22, photo 1, fig. 6, map (1978). Type: Mozambique, Boka, lower Buzi R. bank, *Swynnerton* 1148 (BM!, holo., K!, iso.)

Shrub 1–3 m high; branches pale brown, brittle. Leaves petiolate; blade narrowly ovate, 4–12 cm long, 1–4 cm wide, acute to acuminate at the apex, cuneate at the base, glabrous; petiole 4–11 mm long. Inflorescence solitary, few-flowered; peduncle 4–23 mm long with a few distal bracts; bracts oblong, 1–4 mm long, 1–2 mm wide; pedicels 7–19 mm long. Flowers with sepals persistent, 4–9 mm long; corolla white, tube 13–21 mm long, cylindrical; lobes obtriangular, oblique, with two apices, one rounded and one acute, 8–18 mm long, 7–15 mm wide; stamens inserted about halfway up the tube; pistil 13–18 mm long. Fruits probably orange, the mericarps syncarp for 66–90%, 13–23 mm long, 15–23 mm in diameter, not ridged, not beaked; seeds 5–7 mm long, 2.5–3.5 mm in diameter. Fig. 16 (p. 51).

TANZANIA. Uzaramo District: Pugu Hills Forest Reserve, 25 May 1994, *Mwasumbi* 17897!
DISTR. **T** 6; Somalia, Mozambique, South Africa
HAB. Dry coastal evergreen forest; 230 m
USES. None recorded

SYN. *Ephippiocarpa orientalis* (S. Moore) Markgr. in Notizbl. Bot. Gart. Mus. Berl. 8 (74): 310 (1923)
　　Conopharyngia humilis Chiov. in Atti Soc. Nat. Mat. Modena 66: 10 (1935). Type: Somalia, between Mogadishu and Balad, Damero, 15 Mar. 1933, *Chiovenda* s.n. (FT!, holo.)
　　Ephippiocarpa humilis (Chiov.) Boiteau in Adansonia ser. 2, 16 (2): 280 (1976)

14. CARVALHOA

K. Schum. in E. & P. Pf. 4, 2: 189 (1895); Leeuwenberg in F.Z. 7, 2: 444 (1985) & in Agric. Univ. Wageningen Papers 85 (2): 49–55 (1985)

Shrub repeatedly dichotomously branched; white latex present in most parts. Leaves opposite; petiole or leaf bases of a pair united at the base, forming a very short ochrea, with a single row of colleters in the axils. Inflorescence in pairs in the forks, pedunculate, corymbose, lax. Corolla white, creamy or pale yellow, with many red longitudinal lines at the base of the lobes and at the apex of the tube; lobes in bud overlapping to the left; stamens included; anthers introrse, sessile, narrowly triangular, acuminate, apex sterile, sagittate at the base; pistil glabrous; ovary broadly ovoid, of 2 free carpels, surrounded at the base by an entire disk; pistil head composed of a stigmatic subglobose basal part surrounded by an entire, thin, slightly recurved ring and a bifid stigmoid apex. Fruit of 2, free pod-like follicles, dehiscent adaxially; pericarp soft, orange inside; seeds surrounded by a darker orange pulpy aril, obliquely ellipsoid or ovoid; endosperm copious, starchy, creamy, ruminate, surrounding the spatulate embryo.

A monotypic genus in eastern and SE Africa.

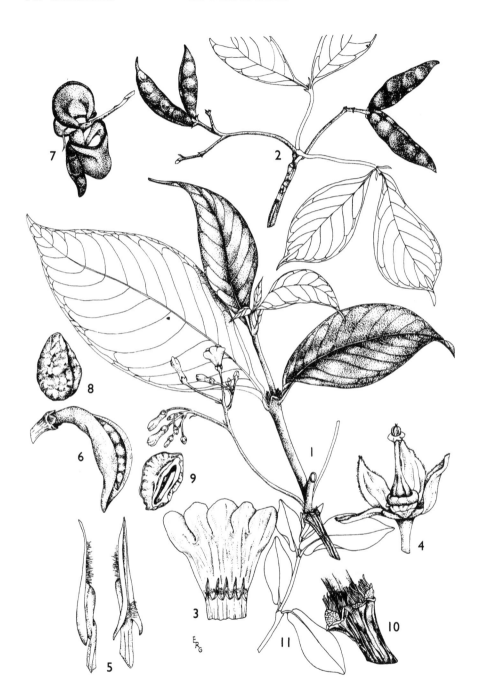

FIG. 17. *CARVALHOA CAMPANULATA* — **1**, flowering branch, × ¹/₂; **2**, fruiting branch, × ¹/₂; **3**, opened corolla, × 3; **4**, calyx with pistil, × 5; **5**, stamens, × 8; **6**, dehiscent fruit, × 1; **7**, open fruit, × ¹/₂; **8**, seed, × 3¹/₂; **9**, LS of seed, × 3¹/₂; **10**, node with colleters, × 3; **11**, branchlet, × ¹/₂. 1 from *Goetze* 1343; 2 from *Baagoe et al.* 93; 3–5 from *Schlieben* 5454a; 6 from *Drummond & Hemsley* 1703; 7–10 from *Milne-Redhead & Taylor* 935; 11 from *Mendonça* 1136. From Meded. Landbouwhogeschool 85(2), drawn by E.Riemer-Gerhardt, reproduced by permission.

C. campanulata *K. Schum.* in E. & P. Pf. 4, 2: 189 (1895); Leeuwenberg in F.Z. 7, 2: 445, t.102 (1985) & in Agric. Univ. Wageningen Papers 85 (2): 50, t. 1 (1985); K.T.S.L.: 479, ill. (1994). Type: Mozambique, Niassa, Mossuril, Cabaceira Pequeña, *Rodriguez de Carvalho* s.n., 1884–1885 (B†, holo. COI, lecto., K, P, Z, iso., chosen by Leeuwenberg)

Shrub 1–5 m high with white latex; branches pale grey-brown, lenticellate; branchlets glabrous or pubescent, lenticellate. Leaves petiolate; blade elliptic, 4–26 cm long, 1.5–13 cm wide, acuminate at the apex, cuneate, rounded or subcordate at the base, membraneous when dry, glabrous, rarely pubescent; petiole 0–7 mm long, glabrous, rarely pubescent. Inflorescence 4–17 cm long, 3–10 cm wide, glabrous in all parts, rarely pubescent; peduncle slender, 1.5–7 cm long; bracts minute, sepal-like, early caducous, leaving large scars; pedicels thin, 5–20 mm long. Flowers with sepals persistent, up to 3.5 mm long; corolla tube 8–10 mm long, campanulate; lobes suborbicular, 3–6 mm long, 3.5–6 mm wide, rounded, not forming a head in mature bud; stamens inserted 2.8–3.5 mm above the corolla base; pistil 4.5–5 mm long; style split at the base, 1.2–1.5 mm long; ovules 30 in each carpel. Fruits yellow or pale orange, recurved or straight, 3–6 cm long, 0.8–1 cm wide, acuminate at the apex, indented around the seeds when dry; seeds 5.5–6 mm long. Fig. 17 (p. 53).

KENYA. Kwale District: Longomwagandi Forest, 17 Mar. 1991, *Luke & Robertson* 2727! & Mwele Mdogo Forest, 28 May 1987, *Robertson* 4673! & Makadara Forest, 17 Jan. 1989, *Omino* 11!
TANZANIA. Lushoto District: Amani, 29 Jan. 1940, *Greenway* 5915!; Morogoro District: Bunduki Forest, Mar. 1953, *Paulo* 55!; Iringa District: Kigogo Forest Reserve, 18 Dec. 1961, *Richards* 15747!
DISTR. **K** 7; **T** 3, 6–8; Malawi and Mozambique
HAB. Moist forest, riverine forest; 300–2300 m
USES. Roots as minor medicine

SYN. *C. macrophylla* K. Schum. in E.J. 30: 381 (1901); T.T.C.L.: 48 (1949). Type: Tanzania, Rungwe/Mbeya District, Ngurumbi [Kurumbi] Mt, *Goetze* 1343 (B†, holo., BR, lecto, E, K, Z, isolecto., erroneously as 1143, chosen by Leeuwenberg)
   C. petiolata K. Schum. in E.J. 33: 317 (1903). Type: Tanzania, Lushoto District, Usambara Mts, Derema, *Scheffler* 219 (B†, holo., PRE, lecto., chosen by Leeuwenberg)

15. SCHIZOZYGIA

Baill. in Bull. Soc. Linn. Paris 1: 752 (1888); Barink in Meded.
Landbouwhogeschool 83 (7): 47–51 (1983)

Shrub, repeatedly dichotomously branched, with white latex. Leaves opposite, petiolate; colleters present in 1–2 rows in leaf axils. Inflorescence two together in forks of branches, congested. Flowers 5-merous. Sepals free, subequal, imbricate. Corolla salver-shaped, lobes in bud overlapping to the right; stamens included, anthers sessile, introrse, upper two thirds fertile and dehiscent longitudinally; ovary of two free rounded carpels; pistil head with a basal cylindrical part 0.4–0.9 mm long and a bifid stigmoid apex; disk present, adnate to the ovary. Fruits of two ellipsoid almost free follicles, laterally compressed. Seeds surrounded by a thin pulpy red or orange aril; endosperm white, copious; embryo white, spatulate.

A monotypic genus in Central and East Africa and Comoro Islands.

S. coffaeoides *Baill.* in Bull. Soc. Linn. Paris 1: 752 (1888) & Hist. Pl. 10: 202 (1889); K. Schum. in E. & P. Pf. 4, 2: 109 (1895); U.O.P.Z.: 444 (1949); T.T.C.L.: 55 (1949); K.T.S.: 49 (1961); Verdcourt & Trump, Comm. Poison. Pl. E. Afr.: 132 (1969); Barink in Meded. Landbouwhogeschool 83 (7): 49, t. 7 (1983) & in F.Z. 7, 2: 447, t. 103 (1985); K.T.S.L.: 485, ill., map (1994). Type: Tanzania, Zanzibar, annis 1847–1852, *Boivin* s.n. (P, holo., BM, BR, iso.)

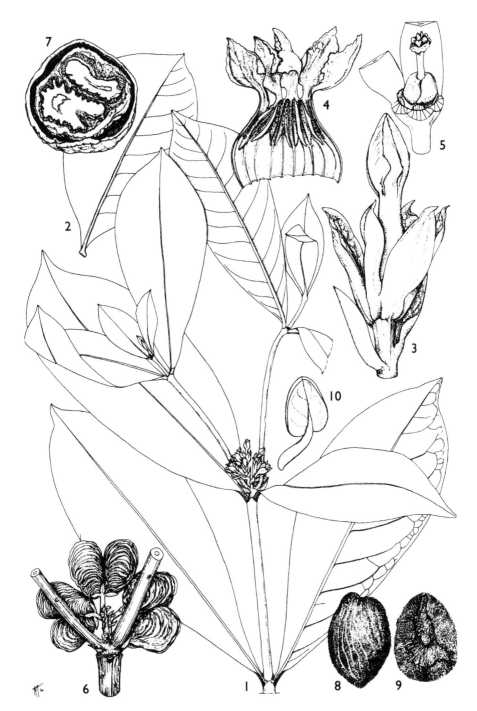

FIG. 18. *SCHIZOZYGIA COFFAEOIDES* — **1**, flowering branch, × ²/₃; **2**, leaf, × ²/₃; **3**, flower, × 6; **4**, opened corolla, × 6; **5**, part of calyx with pistil, × 6; **6**, fruits, × 7; **7**, TS fruit, × 4; **8**, **9**, detail seeds both sides, × 4; **10**, embryo × 6. 1,3–5 from *Peter* 58268, *Stolz* 1693; 2,7 from *Boivin* s.n. 1847–1852; 6, 8–10 from *Bamps* 6301. From Meded. Landbouwhogeschool 83(7), drawn by Y.F. Tan, reproduced by permission.

Shrub 1–4(–8) m high; wood soft, pale yellow; bark rough, brown, with pale lenticels, red inside; branchlets yellow to dark brown, glabrous. Leaves petiolate; blade obovate, 2.4–25 cm long, 1.1–11 cm wide, acuminate at the apex, cuneate at the base, glabrous; petiole 0.5–9 mm long, glabrous. Inflorescence 7–15 mm long, glabrous in all parts; peduncle up to 3 mm long, glabrous; bracts narrowly oblong, 3–5 mm long, acute, glabrous; pedicels 2–3 mm long. Flowers fragrant; sepals elliptic, 3.1–6 mm long, 1.5–3.9 mm wide, acute or acuminate, glabrous, with 5–10 colleters near the base; corolla white or pale yellow, 6.5–9.5 mm long; tube yellow, cylindrical or urceolate, 4–5.2 mm long, glabrous outside, pilose inside around the anthers; lobes creamy to yellow, spreading, obliquely obovate to nearly hook-shaped, 2.5–4 mm long, 2–4 mm wide, glabrous; pistil 2.5–5 mm long; ovules 8–15 in each carpel; style 1–3 mm long, glabrous. Fruit yellow to orange, 7–15 mm long, 5–10 mm in diameter, dehiscent, glabrous; pericarp thinly coriaceous, irregularly striate, grooved when dry; seeds dark brown, obliquely ellipsoid, 5–6 mm long, with a deep groove at the middle of the hilar side. Fig. 18 (p. 55).

Kenya. Kwale District: Shimba Hills, Pengo forest, 11 Feb 1953, *Drummond & Hemsley* 1206!; Kilifi District: Kambe Rocks, 9 Jul. 1987, *Luke & Robertson* 469!; Tana River District: Tana R., *Collinson* 40!

Tanzania. Lushoto District: E Usambaras, Sigi valley, 29 Dec. 1956, *Verdcourt* 1747!; Morogoro District: Kanga Mts S Ridge, *Pocs* 6137/0!; Iringa District: Mohasi R., Mar. 1954, *Carmichael* 430; Zanzibar, Masingini Forest, May 1981, *Ruffo* 1662

Distr. **K** 7; **T** 3, 6–8; **Z**, **P**; Congo (Kinshasa), Somalia, Angola, Malawi and Comoro Islands

Hab. Moist forest and riverine forest; 0–1100 m

Uses. Minor medicine against inflammation and dizziness; roots for skin diseases; fruits are said to be very poisonous

16. DIPLORHYNCHUS

Fic. & Hiern in Trans. Linn. Soc., Bot. 2, 2: 22 (1881); Plaizier in Meded. Landbouwhogeschool 80 (12): 28–35 (1980)

Trees or shrubs, sometimes lianas, with white or yellow sticky latex. Leaves (sub-) opposite. Inflorescence terminal or in the axils of upper leaves, thyrsoid, lax to congested. Flowers fragrant, actinomorphic except for the slightly unequal sepals. Corolla salver-shaped; tube with a subcampanulate upper and a cylindrical narrower basal portion, constricted at the throat; lobes overlapping to the left, with corona at the base; stamens included; anthers completely fertile; ovary of two free carpels, coherent at the base, placentation parietal, ovules 4 in each carpel; pistil head with a pubescent, basal stigmatic subcylindrical part and a bifid stigmoid apex. Fruits composed of two woody follicles, coherent at the base, 4-seeded. Seeds compressed, obliquely oblong, dorsal side straight, ventral side abruptly curved, long-winged at apex.

Monotypic genus restricted to Africa.

D. condylocarpon (*Muell. Arg.*) *Pichon* in Mém. Mus. Nat. Hist. Nat. sér. nov. 19: 368 (1948); Codd in F.S.A. 26: 265, t. 38 (1963); Verdcourt & Trump, Comm. Poison. Pl. E. Afr.: 130 (1969); Plaizier in Meded. Landbouwhogeschool 80 (12): 28, t. 6 (1980) & in F.Z. 7, 2: 454, t. 106 (1985). Type: Africa, probably Angola, 'Quirengue', collector unknown (P, holo.)

Tree or shrub 1–12(–20) m tall; trunk 1–20 cm in diameter; bark smooth to rough, longitudinally fissured or reticulate, greyish-black to brown; branchlets drooping, puberulent to glabrous. Leaves petiolate; blade ovate or obovate, elliptic to suborbicular, 2.6–12 cm long, 1.1–6.7 cm wide, acuminate to rounded to

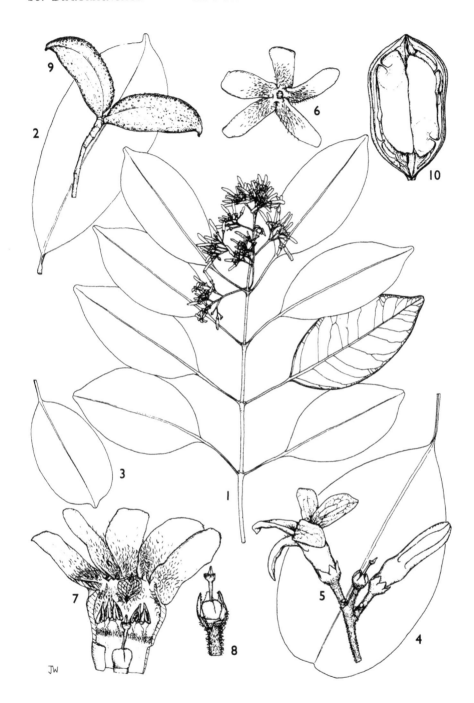

FIG. 19. *DIPLORHYNCHUS CONDYLOCARPON* — **1**, habit, × ²/₃; **2–4**, leaf, × ²/₃; **5**, flower, × ²/₃; **6**, petals seen from above × 4; **7**, anthers and pistil × 6; **8**, pistil, × 6; **9**, fruit × ²/₃; **10**, fruit open with two seeds × ²/₃. 1, 6–8 from *Norgramm* 248; 2, 9 from *Brass* 17414; 3 from *Swynnerton* 39; 4, 10 from *Schlieben* 3036; 5 from *Simon & Ngoni* 1299. From Meded. Landbouwhogeschool 80(12), drawn by J. Williamson, reproduced by permission.

emarginate or mucronate at the apex, cuneate to obtuse at the base, glabrous to pubescent; secondary veins conspicuous, in 6–14(–19) pairs, sometimes with tufts of glandular hairs in the axils (domatia?); petiole 0.5–3.7 cm long, puberulent to glabrescent, with or without glands. Inflorescence 1.5–14 cm long, glabrous to pubescent, sometimes with glandular hairs; peduncle 0.6–4.5 cm long; pedicels 0.5–3 mm long. Flowers with sepals ovate, connate at the base, 0.5–1.5 mm long, 0.25–0.75 mm wide, pubescent to tomentose outside; corolla white to creamy, rarely reddish to orange, tube 1.5–3.2 mm long, 1–1.8 mm wide, glabrous to slightly puberulous outside, velutinous inside in the upper part; lobes narrowly oblong or obovate, pilose inside towards the base, with many glandular hairs, 3.5–6 mm long, 0.7–2 mm wide, between the bases of all lobes a glabrous scale 0.5–1.2 mm long; pistil 1.2–2.4 mm long; style filiform, 0.5–1.3 mm long. Fruits obliquely oblong, 2.2–6.6 cm long, 1.1–2.2 cm in diameter; seed 2.5–5.5 cm long, wing 2–3.2 cm long, 1–1.9 cm wide. Fig. 19 (p. 57).

TANZANIA. Tabora District: Itulu Hill, 9.5 km E of Lyela, Sept. 1951, *Groome* 21!; Kilosa Dsitrict: Kilosa road boundary, 32 km from HQ, 22 Jun. 1973, *Greenway & Kanuri* 15210!; Kilwa District: 12 km from Njinjo on road to Kilwa, 15 Oct. 1978, *Magogo & Rose Innes* 369!
DISTR. **T** 1, 3–8; Angola, Congo (Kinshasa), Malawi, Mozambique, Zimbabwe, Botswana, Namibia and South Africa
HAB. Woodland in rocky sites, wooded grassland and forest margins; 500–1400 m
USES. Latex used for bird lime; roots and fruits used as minor medicinals; wood used for arrows

SYN. *Aspidosperma condylocarpon* Muell. Arg. in Martius, Fl. Brasil. 6(1): 55 (1860)
   Diplorhynchus mossambicensis Oliv. in Hook., Ic. Pl. ser. 3, 4: 40, t. 1355 (1881); P.O.A. C: 316 (1895); Stapf in F.T.A. 4, 1: 107 (1902); T.T.C.L.: 50 (1949); Duvign. in Bull. Soc. Roy. Bot. Belg. 84: 265 (1952). Lectotype: Malawi, Shire Highlands, *Buchanan* s.n. (K, lecto., chosen by Plaizier)
   Diplorhynchus angustifolia Stapf in F.T.A. 4, 1: 107 (1902); T.T.C.L.: 49 (1949). Lectotype: Tanzania, Ugalla, Kabombue, *Böhm* 29a (Z, lecto., chosen by Plaizier)
   D. condylocarpon (Muell. Arg.) Pichon ssp. *mossambicensis* (Oliv.) Duvign. in B.S.B.B. 84: 265 (1952)
   D. condylocarpon (Muell. Arg.) Pichon ssp. *mossambicensis* var. *mossambicensis* f. *angustifolius* (Stapf) Duvign. in B.S.B.B. 84: 265 (1952)

17. **ALSTONIA**

R. Br. in Mem. Wern. Soc. 1: 75 (1811), nom. conserv.; K. Schum. in E. & P. Pf. 4 (2): 138 (1895); de Jong in Meded. Landbouwhogeschool 79 (13): 2–16 (1979)

Trees or shrubs; branches verticillate, in whorls of 4–5. Leaves verticillate or opposite (not in our area), petiolate, with or without intrapetiolar stipules which are often adnate to the petiole; colleters present in leaf axils. Inflorescence terminal, thyrsoid or compound, subumbellate, in whorls of 1–5. Flowers 5-merous, actinomorphic. Corolla white, yellow or red; tube cylindrical, widened around anthers, thickened at the throat; lobes spreading, overlapping to the left; stamens included; anthers basifixed, introrse; disk present, free or adnate to the ovary, often indistinct; ovary 2-celled, apocarpous or syncarpous, ovules numerous, style terete, stigma bifid, subtended by a cylindrical or slightly conical clavuncula. Fruit of two follicles, free or connate at the base. Seeds numerous, pubescent, ciliate, sometimes winged.

A genus of 40 species in the tropics of Asia and Northern Australia and represented by two species in Africa.

A. venenata R. Br. has been cultivated in Kampala (Makerere University, May 1971, *Lye* 6050) and Kenya (Sotik, Kibajet Estate, Sep. 1962, *Ossent* 696; Nairobi, June 1958, *Bell* H193/58 and Muguga, Dec. 1960, *Greenway* 9736). It is a shrub to 1.5 m with lanceolate leaves to 20 cm long; flowers small and white.

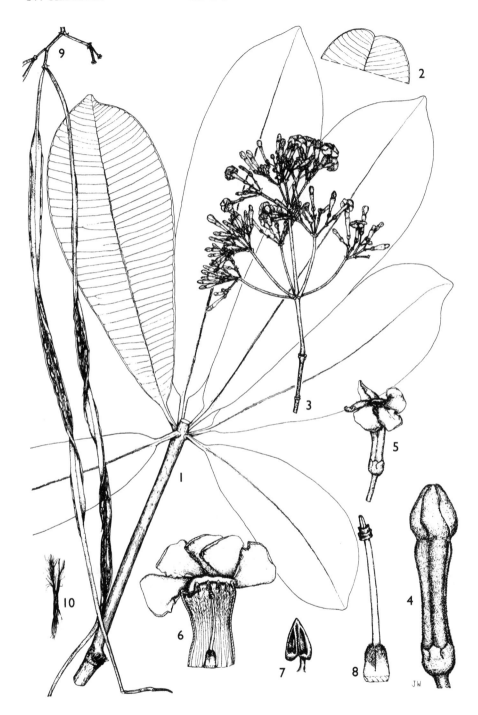

FIG. 20. *ALSTONIA BOONEI* — **1**, branch with leaves, × ¹/₂; **2**, leaf apex, × ¹/₂; **3**, inflorescence, × ¹/₂; **4**, flower bud, × 3; **5**, flower, × 1¹/₂; **6**, flower opened, × 2; **7**, anther, × 10; **8**, pistil × 5; **9**, fruit, × ¹/₂; **10**, seed, × ¹/₅. 1 from *Eggeling* 1547; 2–3 from *P. Wit* 2340; 4–8 from *P. Wit* 2340 spirit collection; 9–10 from *Gille* 84bis. From Meded. Landbouwhogeschool 79(13), drawn by J. Williamson, reproduced by permission.

A. boonei *de Wild.* in F.R. 13: 382 (1914) & in B.J.B.B. 5: 405 (1919); I.T.U. ed. 2: 24, t. 5 (1952); Huber in F.W.T.A. ed. 2, (2): 68 (1963); De Jong in Meded. Landbouwhogeschool 79 (13): 5, t. 1 (1979); Hamilton, Field Guide Uganda For. Trees: 163 (1981). Type: Congo (Kinshasa), Uele, Nala, *Boone* 2 (BR, holo.)

Tree 25–30(–40) m high with white latex; trunk up to 1.4 m in diameter, with or without buttresses up to 8 m high; bark grey, white or yellowish, smooth or scaly. Leaves in whorls of 4–9, petiolate; blade (narrowly) obovate, rarely oblong, 8–19(–24) cm long, 3.5–7.5 cm wide, acuminate, rarely obtuse or retuse at the apex, decurrent into the petiole, coriaceous; secondary veins in 25–50(–60) pairs, forming an angle of almost 90° with the midrib and a neat submarginal vein; petiole 0.6–2.9 cm long. Inflorescence compound-subumbellate, 7.5–23 cm long, pubescent in all parts; bracts small, sepal-like; pedicel 0.2–0.6 cm long. Flowers with sepals broadly ovate, 1–2.2 mm long, 1–2.7 mm wide; corolla yellow-green to pale green; lobes pale yellow, rarely white, salver-shaped; tube 6.5–14 mm long; lobes obliquely ovate or obovate, 2.5–6.5 mm long, acute or obtuse; pistil 4.5–11.5 mm long. Fruits of two linear follicles, 22–57 cm long, 0.2–0.4 cm wide, pubescent, dehiscent; seeds many, in two rows, with a thickened margin, bearing long stiff hairs at the ends and much shorter ones at the side. Fig. 20 (p. 59).

UGANDA. Bunyoro District: Budongo Forest Reserve, 9 Feb. 1972, *Synnott* 802!, & Feb. 1932, *Harris* 403; Mengo District: Kampala, near BAT Mess, Dec. 1934, *Eggeling* 1547
TANZANIA. Kigoma District: Gombe Stream Reserve, Mitumba Valley, 7 Sep. 1993, *Mbago & Mpongo* 1238 (fide DSM)
DISTR. **U** 2, 4; **T** 4; from Senegal to Ethiopia to Congo (Kinshasa)
HAB. Rainforest, ground-water or riverine forest; 800–1200 m
USES. Minor timber with soft wood, prized for making bowls

SYN. [*A. congensis* sensu Stapf in F.T.A. 4, 1: 121 (1902); F.W.T.A. ed.1, 2 (1): 42 (1931); Eggeling & Harris, 15 Uganda Timber Tr.: 9 (1939), *non* Engl.]

18. RAUVOLFIA

L., Sp. Pl. 1: 208 (1753); van Dilst & Leeuwenberg in B.J.B.B. 61: 21–69 (1991)

Shrubs or trees, rarely rhizomatous undershrubs (but not in our area), with white latex in all parts, often candelabrum-shaped; branches in whorls of 3–5; branchlets glabrous to lanate. Leaves in whorls of 3–6, sometimes opposite on lower nodes, often confined to apices of branchlets, with numerous colleters in leaf axils; blade glabrous. Inflorescence terminal, rarely axillary, umbellate, lax or congested, few–many-flowered. Flowers 5-merous, fragrant. Sepals connate at the base, slightly unequal. Corolla variously shaped; lobes in bud overlapping to the left; stamens included or exserted; filaments short; anthers mucronate or apiculate at the apex, cordate at the base, glabrous; ovary superior of two free to syncarpous carpels, glabrous; disk annular or cupular, entire, undulate or slightly crenate, glabrous; pistil head cylindrical, with a basal collar and bifid stigmoid apex. Fruit of two apocarpous or partly to completely syncarpous drupes each with a single stone. Seeds laterally compressed, obliquely ovoid or ellipsoid, with a large embryo; embryo surrounded by a thin endosperm.

A pantropical genus of ± 60 species, 7 of which are represented in continental Africa.

1. Corolla lobes elliptic, usually > 5 mm long; stamens inserted in the lower half of the corolla tube; inflorescence lax, 1–11-flowered · *4. R. volkensii*
Corolla lobes ovate, obovate or dolabriform, < 5 mm long; stamens mostly inserted in upper half of corolla tube (half-way in *R. caffra*); inflorescence congested or lax, usually more than 20-flowered · 2

2. At least some branches of inflorescence puberulous · · · ·   5. *R. vomitoria*
   Inflorescence branches glabrous · 3
3. Inflorescence < 7 cm long, less than 50-flowered; leaf acumen
   usually > 5 mm long ·   2. *R. mannii*
   Inflorescence 5–15 cm long, more than 50-flowered; leaf
   acumen < 5 mm long · 4
4. Pedicels < 2 mm long; corolla tube densely setose at mouth
   and base of lobes; fruit 10–30 mm long · · · · · · · · · · ·   1. *R. caffra*
   Pedicels > 2 mm long; corolla tube sparsely pubescent at
   the mouth; fruit 5–10 mm long · · · · · · · · · · · · · · ·   3. *R. mombasiana*

1. **R. caffra** *Sond.* in Linnaea 23: 77 (1850); I.T.U. ed. 2: 30 (1952); K.T.S.: 48 (1961); Huber in F.W.T.A. ed. 2, 2: 69 (1963); Codd in F.S.A. 26: 275 (1963); de Kruif in F.Z. 7 (2): 460, t. 108 (1985); van Dilst & Leeuwenberg in B.J.B.B. 61: 24, t. 1 (1991); U.K.W.F. ed. 2: 172 (1994); K.T.S.L.: 483, ill., map (1994). Type: South Africa, southern slopes of Magaliesberg, *Zeyher* 1183 (S, holo., AWH, BM, BP, BOL, G, K, P, W, iso.)

Shrub or tree 2–40 m high; trunk 5–100 cm in diameter; bark smooth and lenticellate or rough and corky, longitudinally fissured, grey to brown; wood soft; branches and branchlets brown, lenticellate, often 4–5-angled or -winged, with conspicuous leaf scars. Leaves in whorls of 3–6, confined to the apices of branches, petiolate; blade narrowly obovate to elliptic, 1.8–60(–70) cm long, 0.8–15(–19) cm wide, subacute to acute at the apex, cuneate or attenuate at the base, sometimes decurrent into the petiole, glossy, glabrous; 12–35(–45) pairs of secondary veins; petiole 0–60 mm long, glabrous. Inflorescence congested, on ultimate branches only, up to 450-flowered; peduncle 5.8–13.5 cm long, glabrous; pedicels 0.1–2 mm long, glabrous. Flowers fragrant; sepals ovate, 0.5–1.3 mm long, 0.3–1.5 mm wide, obtuse at the apex, glabrous, sometimes slightly ciliate; corolla hypocrateriform, (greenish) white; tube broadly cylindrical, 3.2–5.5 mm long, densely setose in the mouth and at base of lobes; lobes ovate or obovate, 0.6–1.6 mm long, 0.8–1.6 mm wide; stamens inserted 1.9–4.2 mm above the corolla base; disk 0.2–0.7 mm high, annular or cupular; pistil 1.2–3.8 mm long; ovary of 2 free or partly fused carpels, globose to obovoid, glabrous; style 0.6–3.2 mm longr. Fruit dark red or black, often lenticellate, syncarpous when both carpels develop, obcordate, bilobed, 10–30 mm long, 7–23 mm in diameter; when only one carpel developes, subglobose to ellipsoid and 5–20 mm in diameter; seeds 7–13 mm long. Fig. 21 (p. 62).

UGANDA. Toro District: Bunyangabo, S of Lake Kabaleka, 14 Mar. 1952, *Osmaston* 1329!; Mbale District: Sabei, near Mt Elgon, 18 Nov. 1926, *Snowden* 1038!; Mengo District: Namanue forest, Sept. 1932, *Eggeling* 534!
KENYA. Masai District: Masai Mara Game Reserve, E of Oloitikoishi, 24 Jun. 1979, *Kuchar* 11579!; Meru District: N bank of Kindani R. near Kindani, 1 May 1972, *Ament & Magogo* 72!; Trans Nzoia District: Kitale Musuem, 3 Mar. 1986, *Ekkens* 460!
TANZANIA. Arusha Disrict: Arusha, Dec. 1949, *Bally* 7667!; Morogoro District: Mtibwa Forest Reserve, Aug. 1952, *Semsei* 879!; Iringa District: Ihongali, Nov. 1953, *Carmichael* 318!
DISTR. **U** 1–4; **K** 1, 3–7; **T** 1–8; widely distributed in tropical Africa from Togo to Angola in the W and from Sudan to South Africa
HAB. Riverine forest, swamp forest, moist forest; 450–1950 m
USES. Wood soft, used for building; also cultivated in Nairobi

SYN. *R. natalensis* Sond. in Linnaea 23: 78 (1850); Stapf in F.T.A. 4, 1: 111 (1902); T.T.C.L.: 54 (1949). Type: South Africa, Durban [Port Natal], *Gueinzius* 183 (S, holo., G, W, iso.)
  R. inebrians K. Schum. in Engl., P.O.A. A: 93, B: 352, C: 318 (1895); Stapf in F.T.A. 4, 1: 112 (1902); T.T.C.L.: 54 (1949). Lectotype: Tanzania, Kilimanjaro, Marangu, *Volkens* 1415 (BM, lecto., E, G, K isolecto., chosen by van Dilst & Leeuwenberg)
  R. ochrosiodes K. Schum. in P.O.A. C: 318 (1895); Stapf in F.T.A. 4, 1: 111 (1902); T.T.C.L.: 54 (1949). Type: Tanzania, ?Mara District, Uthiri [Itarige], *Fisher* 377 (K, holo.)

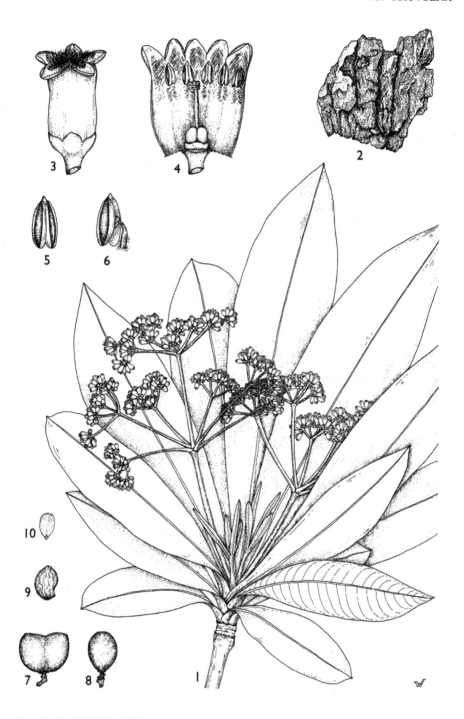

FIG. 21. *RAUVOLFIA CAFFRA* — **1**, habit, × ²/₃; **2**, bark, × ²/₃; **3**, flower, × 6; **4**, part of corolla opened out, × 6; **5**, stamen, adaxial side, × 14; **6**, stamen, lateral view, × 14; **7**, **8**, fruits × ²/₃; **9**, seed, × ²/₃; **10**, embryo, × ²/₃. 1 from *Leeuwenberg* 10810; 2 from *Leeuwenberg* 1256; 3–6 from *Leeuwenberg* 10868; 7–10 from *Leeuwenberg* 10873. From F.Z., drawn by W.Wessels, reproduced by permission.

R. obliquinervis Stapf in F.T.A. 4, 1: 112 (1902); T.T.C.L.: 54 (1949). Type: Tanzania, Lushoto District, Usambara, Handeni, Kizara, *Holst* 2360 (K, holo.)

R. goetzei Stapf in F.T.A. 4, 1: 113 (1902); T.T.C.L.: 54 (1949). Type: Tanzania, Morogoro District, Ruhambe, *Goetze* 385 (K, holo., BM, E, iso.)

R. oxyphylla Stapf in K.B. 1908: 407 (1908); I.T.U. ed. 2: 30 (1952); Hamilton, Field Guide Uganda For. Trees: 163 (1981). Lectotype: Uganda, Mbale District, plain below Ruwenzori, *Dawe* 603 (K, lecto., chosen by van Dilst & Leeuwenberg)

2. **R. mannii** *Stapf* in K.B. 1894: 21 (1894); Huber in F.W.T.A. ed. 2, 2: 69 (1963); de Kruif in F.Z. 7, 2: 459 (1985); van Dilst & Leeuwenberg in B.J.B.B. 61: 38, t. 7 (1991); U.K.W.F. ed. 2: 171 (1994); K.T.S.L.: 483, map (1994). Type: Gabon, Cristal Mts, *Mann* 1720 (K, holo., P, iso.)

Shrub or small tree, 0.3–8 m high; trunk 0.5–2.5 cm in diameter or more; bark scaly, peeling, greenish to greyish brown, lenticellate; branches brown, lenticellate. Leaves in whorls of 3–6, opposite on some nodes; petiolate; blade elliptic or slightly obovate, 2.5–28 cm long, 0.7–10.5 cm wide, acuminate or cuspidate at the apex, acumen 2–35 mm long, cuneate at the base, glabrous; with 6–19 pairs of secondary veins; petiole 1–27 mm long, glabrous. Inflorescence terminal or pseudoaxillary, 3–50-flowered, glabrous in all parts; peduncle 0.2–6 cm long; pedicels 1–8 mm long. Flowers white, lobes pale pink, finely speckled dark red along mid vein; sepals ovate–elliptic, 0.6–2.8 mm long, glabrous; corolla hypocrateriform; tube 2.5–10.5 mm long, glabrous outside; lobes ovate to broadly ovate, dolabriform, often curved to the left, 0.8–3.6 mm long, 0.6–2.4 mm wide, obtuse to rounded at the apex; stamens inserted 2–8 mm from the corolla base; disk cupular, 0.4–0.8 mm high; pistil 1.8–6.5 mm long; ovary globose to ellipsoid; style 0.8–6 mm long; pistil head 0.3–1.2 mm long. Fruits red or purple-black, often laterally compressed, when apocarpous usually one carpel developing, ovoid to almost ellipsoid, 5–12 mm long, 4–7 mm in diameter; when syncapous obcordate, 5–11 mm long, 4–14 mm wide; seeds 4–11 mm long.

UGANDA. Kigezi District: Impenetrable Forest, Sep. 1936, *Eggeling* 3310 & Kinkizi, on slopes of E Isasha Gorge, 5 Aug. 1971, *Katende* 1260!

KENYA. Meru District: 3 km from Maua on Meru–Maua road, 6 Sept. 1967, *Hanid & Kiniaruh* 972!; Kitui District: Mutito Hill, Jan. 1937, *Gardner* 1068!; Teita District: Taita Hills, base of Ngangao Forest, 7 Jul. 1969, *Faden & Evans* 69/876!

TANZANIA. Lushoto District: Shibangeda, near Amani, 30 Jun. 1970, *Kabuye* 199!; Morogoro District: Bunduki Forest Reserve, Mar. 1953, *Paulo* 56!; Iringa District: Image Mt, Nov. 1959, *Procter* 1548!

DISTR. **U** 2; **K** 4, 7; **T** 2, 3, 6–8; Liberia, Ivory Coast, Ghana, Nigeria, Cameroon, Gabon, São Tomé, Congo (Brazzaville), Congo (Kinshasa), Rwanda, Burundi, Angola and Malawi

HAB. Moist forest; 300–2250 m

USES. None recorded

SYN. *R. obscura* K. Schum. in E. & P. Pf. 4 (2): 154 (1895); Stapf in F.T.A. 4, 1: 117 (1904). Lectotype: Congo (Kinshasa), Mukenge, *Pogge* 1080 (K, lecto., HBG, M, isolecto., chosen by van Dilst & Leeuwenberg)

R. rosea K. Schum. in P.O.A. C: 317 (1895); T.T.C.L.: 55 (1949); K.T.S.: 49 (1961). Type: Tanzania, Lushoto District, Usambara, Lutindi, *Holst* 3250 (K, holo., HBG, M, iso.)

R. longiacuminata De Wild. & Durand in B.S.B.B. 38 (1): 205 (1899), as *longeacuminata*. Type: Congo (Kinshasa), Bas-Congo, *Cabra* 19 (BR, holo.)

NOTE. Flowers are variable in size, shape and colour.

3. **R. mombasiana** *Stapf* in K.B. 1894: 21 (1894); U.O.P.Z.: 431 (1949); T.T.C.L.: 54 (1949); K.T.S.: 48 (1961); de Kruif in F.Z. 7, 2: 462 (1985); van Dilst & Leeuwenberg in B.J.B.B. 61: 48, t. 11 (1991); K.T.S.L.: 484, map (1994). Lectotype: Kenya, Mombasa, *Hildebrandt* 2011 (BM, lecto., K, isolecto., chosen by van Dilst & Leeuwenberg)

Shrub or small tree, 0.5–9 m high; trunk 1–10 cm in diameter; bark grey, smooth or rough, lenticellate. Leaves in whorls of 3–6, opposite on some nodes; blade obovate to elliptic or narrowly elliptic, 2.5–25 cm long, 1–7 cm wide, acute to shortly acuminate at the apex, cuneate or attenuate at the base, glabrous, with 8–25 pairs of secondary veins. Inflorescence in whorls of 1–3 many-flowered cymes, glabrous in all parts; peduncle 2–15.5 cm long; pedicels 3–8 mm long. Flowers with sepals ovate, 0.6–1.5 mm long, 0.8–1.2 mm wide, acute at the apex, glabrous; corolla hypocrateriform; tube whitish yellow or greenish white, 4–7.6 mm long, sparsely pubescent at the mouth; lobes white, yellowish or greenish white, dolabriform, 0.9–1.6 mm long, 1–2 mm wide, glabrous; stamens inserted at 3–6.8 mm from the base of the corolla tube; disk 0.4–1 mm high, cupular; pistil 2.6–6.1 mm long; style 2–4.6 mm long; ovary obovoid, of 2 free carpels; pistil head cylindrical, 0.4–1 mm long; stigmoid apex 0.1 mm. Fruit orange to red, apocarpous, with 1 or 2 mericarps developing; mericarps ellipsoid, 5–10 mm long, 4–8 mm in diameter; seeds 3–8 mm long.

KENYA. Lamu District: Boni Forest, Marereni, Dec. 1946, *Joy Adamson* 300 in *Bally* 5991!; Kwale District: Shimba Hills, near Pengo Hill, 9 Jan. 1970, *Katende & Lye* 4809!; Mombasa District: N of Mombasa at White Sands, 9 May 1956, *Bally* 10534!
TANZANIA. Mwanza District: Rubondo I. Game Reserve, 9 May 1975, *Ludanga* 2037!; Tanga District: 4 miles W of Tanga, 1 Aug. 1953, *Drummond & Hemsley* 3594!; Kigoma District: Kakombe valley, Kakombe Stream Reserve, 2 Jan 1972, *Parnell* s.n.! Zanzibar, Jozani, Nov. 1930, *Greenway* 2589
DISTR. **K** 7; **T** 1, 3, 6–8; **Z, P**; Mozambique
HAB. Riverine forest, coastal forest or thicket; 0–550(–1800) m
USES. Bark used as rat poison; root decoction used to treat malaria

SYN. *R. monopyrena* K. Schum. in P.O.A. C: 318 (1895); Stapf in F.T.A. 4, 1: 115 (1902). Type: Tanzania, Amboni, *Holst* 2801 (K, holo.)

4. **R. volkensii** (*K. Schum.*) *Stapf* in F.T.A. 4, 1: 116 (1904); T.T.C.L.: 55 (1949); van Dilst & Leeuwenberg in B.J.B.B. 61: 57, t. 16 (1991). Lectotype: Tanzania, Lushoto District, Usambara Mts, Mlalo, *Holst* 174 (K, holo., P, iso.)

Shrub or small tree, 1–5 m high with spreading, sometimes drooping branchlets; branches pale or dark brown, with protruding lenticels. Leaves in whorls of 3–5, sometimes opposite on lower nodes; blade (narrowly) elliptic, 1.5–17 cm long, 0.5–5.5 cm wide, acuminate, apiculate or caudate at the apex, acumen 5–19 mm long, glabrous; petiole (1) 2–18 mm long, glabrous. Inflorescence 1–11-flowered; peduncle 0.4–7 cm long, glabrous; bracts scaly; pedicels (1.5–)3–10 mm long, glabrous. Flowers white, fragrant or not; sepals green, ovate or subulate, 1.3–2.5 mm long, acuminate at the apex, glabrous; corolla hypocrateriform, tinged pink in bud; tube (5–)7.2–17.5 mm long, glabrous outside, pubescent at the insertion of stamens and villose to the mouth; lobes elliptic, (3–)5.8–13 mm long, (1.5–)3.2–6.8 mm wide, obtuse or rounded at the apex, glabrous; stamens inserted (3.2–)4.5–5.8 mm above the corolla base; disk 0.5–0.9 mm high, annular or cupular; pistil (3.3–)4.7–5.8 mm long; ovary with partly fused carpels; style (1.5–)3.2–3.5 mm long. Fruits bright orange to red, laterally compressed; when syncarpous, obcordate, 7–10 mm long, 11–19 mm in diameter, when one carpel developing obliquely ellipsoid, 10–13 mm long, 5–7 mm in diameter; seed 4–6 mm long, 3–5 mm in diameter.

TANZANIA. Moshi District: Rombo Forest Reserve, 21 Nov. 1983, *Magogo* 2494!; Pare District: Chome Forest Reserve, 2 Feb. 1953, *Parry* 197!; Lushoto District: German bridge, Magamba, 11 Feb. 1964, *Rajabu* EA 12970!
DISTR. **T** 2, 3; endemic to NE Tanzania
HAB. Moist forest or riverine forest; 900–2100 m
USES. None recorded

Syn. *Tabernaemontana volkensii* K. Schum. in P.O.A. C: 316 (1895)
   Rauvolfia faucium Engl., nom. nud. in Pflanzenw. Afr. 1, 1: 299 (1910); T.T.C.L.: 54 (1949)
   R. oreogiton Markgr. in N.B.G.B. 8: 287 (1923); T.T.C.L.: 55 (1949). Type: Tanzania,
   Lushoto District, Usambara Mts, Kwai, *Eick* 323 & 380 & Magamba, *Deninger* 2741 &
   Lushoto, *Heinrich* 2847 (all four B†, syn.)

5. **R. vomitoria** *Afzel* in Stirp. Guin. Med.: 1 (1818); Stapf in F.T.A. 4, 1: 115 (1902);
Huber in F.W.T.A. ed. 2, 2: 69 (1963); van Dilst & Leeuwenbrg in B.J.B.B. 61: 60, t.
18 (1991). Type: Sierra Leone, *Afzelius* s.n. (UPS, holo., BM, MO, P, iso.)

Shrub or tree 0.5–20(–40) m high; trunk 1–80 cm in diameter; bark smooth,
striate or fissured, light to dark grey-brown or dark brown; wood light yellowish to
white; branches pale to dark brown, or grey-brown, smooth, lenticellate; branchlets
in whorls of 3–5, pale to greenish brown, smooth. Leaves in whorls of 3–5; blade
elliptic or narrowly elliptic, 3.4–27 cm long, 2–9 cm wide, apiculate at the apex,
acumen 1–18 mm long, glabrous, with 8–17 pairs of secondary veins; petiole 6–35 mm
long. Inflorescence in whorls of 1–4, dense, 15–450-flowered, puberulous in all
parts; peduncle 1.5–8.6 cm long; pedicels 1–4.5 mm long. Flowers fragrant; sepals
ovate, 1–2.2 mm long, 0.9–2.1 mm wide, acute at the apex; corolla hypocrateriform,
greenish white to yellow; tube 5.8–10(–12) mm long, glabrous outside, puberulous
inside in 3 small belts; lobes dolabriform, 1.1–2.1 mm long, 1.1–4 mm wide,
glabrous; stamens included, inserted at 4.3–7(–8.6) mm above the corolla base; disk
0.7–2 mm high, cupular; pistil (3.8–)4.2–9.2 mm long; ovary of two partly fused
carpels, ovoid or cylindrical; style (2–)2.4–5.2 mm long; pistil head 0.5–1.2 mm
long, stigmoid apex 0.1 mm long. Fruits bright orange or red, usually 1 mericarp
developing, globose, ovoid or ellipsoid, 8–14 mm long, up to 9 mm in diameter;
seeds ellipsoid, 6–8 mm long.

Uganda. Ankole District: Kashoya-Kitomi Forest, 28 Dec. 1969, *Synnott* 446!; Busoga District:
   Dagussi Landing, 17 Dec. 1954, *Stephens* 112!; Mengo District: 1 km SE of Katamiro Landing,
   27 Sept. 1949, *Dawkins* 416!
Tanzania. Mwanza District: Ukara Is., Feb. 1953, *Smith* 10! & Maisome Is., Lake Victoria, Nov.
   1954, *Carmichael* 457!; Bukoba District: Bukoba town, Apr. 1958, *Procter* 894!
Distr. U 1–4; T 1, 4; widespread in tropical Africa from Senegal to Sudan and South to Angola
Hab. Moist forest and forest margins, thicket; 900–1200 m
Uses. None recorded

Syn. *R. stuhlmannii* K. Schum. in P.O.A. C: 318 (1895). Type: Tanzania, Bukoba District,
   Bukoba, *Stuhlmann* 3621 (B†, holo., K, lecto., chosen by van Dilst & Leeuwenberg)
   R. congolana De Wild. & Durand in B.S.B.B. 39: 34 (1900); Stapf in F.T.A. 4, 1: 115 (1902).
   Type: Congo (Kinshasa), Kiboubou near Kasongwo (Nyangwe), *Dewèvre* 924 (BR, holo.,
   K, iso.)

19. **CERBERA**

L., Sp. Pl.: 208 (1753); Leeuwenberg in Wageningen Agric. Univ. Pap. 98, 3:
5–64 (1999)

Shrubs or trees; white latex present in all parts. Leaves spirally arranged.
Inflorescence terminal or in branching forks. Sepals free or nearly so, imbricate.
Corolla with funnel-shaped tube and oblique lobes. Stamens included, anthers
sessile. Pistil with apex at level of anthers; ovary of 2 separate carpels. Fruit of 2 (or
by abortion 1) mostly separate drupaceous mericarps with hard endocarp. Seeds 1–2
in each mericarp, flat.

A genus of six species, with one in Africa.

FIG. 22. *CERBERA MANGHAS* — **1**, habit, × ²/₃; **2**, flower seen from above, × ²/₃; **3**, opened flower, × ²/₃; **4**, fruits, × ¹/₆. 1–3 from *Rudjiman* 357; 4 from *Rudjiman* 354. From Meded. Landbouwhogeschool 98(3), reproduced by permission.

C. manghas *L.*, Sp. Pl.: 208 (1753), excluding synonyms; U.O.P.Z.: 185, ill. (1949); Leeuwenberg in Taxon 41: 560 (1992) & in Wageningen Agric. Univ. Pap. 98, 3: 21, figs. 2–3, photo 3–7, map (1999). Lectotype: 'East India', *Osbeck* s.n. (LINN 298.2, lecto., chosen by Leeuwenberg)

Shrub or tree 1.5–25 m high; trunk to 70 cm in diameter; bark grey to dark brown, rough, peeling off in clumps, with large lenticels; wood soft, white; glabrous in all parts except for corolla. Leaves petiolate; blade coriaceous, narrowly obovate (elliptic), 5–31 cm long, 1–7(–8) cm wide, apiculate to acuminate (rounded or retuse); 15–40 pairs of secondary veins; petiole 1–4.5 cm long. Inflorescence many-flowered, to 23 cm long, lax near its base but otherwise congested; peduncle 1.5–12 cm long; pedicels 3–28 mm long; bracts sepal-like. Flowers fragrant, usually only one open per inflorescence; sepals pale green, spreading and often recurved, (4–)8–37 mm long, 2–8 mm wide; corolla tube greenish, 17–55 mm long; lobes white, often pinkish outside, obliquely elliptic to obovate, (9–)14–30(–35) mm long, (5–)9–18(–35) mm wide with 5 white or yellow lanate scales 1.5–3 mm long in the centre, pubescent at the mouth and inside the tube; stamens included, anthers subglobose, 1.5–3 mm. Fruit of 2 separate (rarely half-connate) mericarps, purplish red or pale green and suffused with red, mericarps ellipsoid, 5–12 cm long, 3–7 cm in diameter. Fig. 22 (p. 66).

TANZANIA. Pemba: Mikindani, Dec. 1930, *Greenway* 2722; Jundana, June 1928, *Vaughan* 367
DISTR. **P**; Madagascar to Pacific Ocean Islands
HAB. *Pandanus* woodland; near sea level
USES. None recorded

20. **ADENIUM**

Roem. & Schult. in Syst. Veg. 4: 35 (1819), as *Adenum*; Plaizier in Meded.
Landbouwhogeschool 80 (12): 2–26 (1980)

Succulent shrubs or trees 0.2–6 m tall with clear or white latex; lower trunk sometimes up to 1 m in diameter; bark pale grey or brown, smooth; branchlets pubescent at the apex, glabrescent. Leaves alternate, (sub–)sessile, confined to apices of branchlets, with colleters in the axils; stipules minute or absent; blade entire. Inflorescence thyrsoid, lax; peduncle very short or absent. Flowers slightly zygomorphic. Corolla tube infundibuliform to salver-shaped, widened at the throat; lobes obovate, overlapping to the right in bud, with an obcordate corona at the base of and in between the lobes; stamens included to exserted; anthers with long, filiform, usually coherent and twisted appendages at the apex; ovary of two globose carpels, coherent at the base; style cylindrical, split at the base; pistil head subcylindrical; stigmoid apex short, bifid. Fruit of two oblong woody pubescent follicles, coherent at the base, tapering at both ends. Seeds many, with tufts of dirty white to light brown hairs at both ends.

5 species in the tropics of Africa and Southern Arabia.

A. obesum *(Forssk.) Roem. & Schult.* in Syst. Veg. 4 (35): 411 (1819); U.O.P.Z.: 106 (1949); K.T.S.: 44 (1961); Codd in F.S.A. 26: 279 (1963); Verdcourt & Trump, Comm. Poison. Pl. E. Afr.: 129 (1969); Huber in F.W.T.A. ed. 2, 2: 76, t. 217 (1963); Plaizier in Meded. Landbouwhogeschool 80 (12): 13 (1980); Wild Flow. E. Afr.: 142, t. 279 (1987); U.K.W.F. ed. 2: 172 (1994); K.T.S.L.: 475, ill., map (1994). Type: Yemen, Milhan [Melhan], *Forsskål* Herb. 235 (C, holo.)

Shrub or small tree, 2–many-branched, 0.4–6 m high; trunk sometimes bulbous at the extreme base, up to 1 m in diameter (to 2 m elsewhere), bark grey, smooth.

FIG. 23. *ADENIUM OBESUM* — **1**, leafy shoot, × ²⁄₃; **2**, flowering shoot, × ²⁄₃; **3–4**, leaf, × ²⁄₃; **5**, anthers and pistil, × ²⁄₃; **6**, opened flower, × 2; **7, 8**, clavuncula, × 10; **9**, fruit, × ²⁄₃; **10**, seed, × ²⁄₃. 1–2 from *Greenway* 15400, 3 from *Paulo* 1078, 4 from *Khattat* 47, 5–7 from *Leeuwenberg* 10784, 8 from *de Wilde* 8726, 9–10 from *Dekker* 376. From Meded. Landbouwhogeschool 80(12), drawn by J. Williamson, reproduced by permission.

Leaves sessile to sub-petiolate; blade obovate to linear, 3–12(–17) cm long, 0.2–6 cm wide, rounded and mucronate or emarginate at the apex, cuneate at the base, pubescent to glabrous; petiole 0–4 mm long. Inflorescence 1–2.5 cm long, in dense terminal cymes; bracts linear to narrowly oblong, 3–8 mm long; pedicels 5–9 mm long. Flowers with sepals 0.5–1.1 mm long, pubescent; corolla pink to red; tube reddish pink to white suffused with pink, sometimes with red stripes within the throat, 20–50 mm long, 9–17 mm wide, inside with velutinous glandular hairs on the main veins; lobes 9–30 mm long, 5–25 mm wide, apiculate to mucronate; corona 3–5 mm long; stamens barely included or exserted; pistil 11–20 mm long; pistil head 1–1.5 mm long. Fruit grey to pale grey-brown, sometimes fringed with pink, 11–22 cm long, 0.9–2 cm wide; seeds 10–14 mm long, 2–4 mm wide, with tuft of hairs 25–35 mm long. Fig. 23 (p. 68).

UGANDA. Acholi District: Agora Hill, *Jackson* in *Bally* 9840; Karamoja District: Moroto, sine die, *Phillips* in *Bally* 11461 & Kidepo National Park, *Synnott* 1354
KENYA. Meru District: Golo circuit, near Murera River, 18 Dec. 1972, *Ament* 443!; Masai District: Olorgesailie camp site, 3 Aug. 1943, *Bally* 2639!; Kilifi District: 1.5 km S of Jilore, 21 Nov. 1969, *Perdue & Kibuwa* 10026!
TANZANIA. Masai District: Serengeti Plain near Ngorongoro, Oct. 1949, *Princess Windrich-Graetz* B7604! Lushoto District: 9.5 km SW of Mashewa on Mashewa–Magoma road, 8 Jul. 1953, *Drummond & Hemsley* 3216!; Kilosa District: Kilosa, *Greenway & Kanuri* 15400
DISTR. U 1; K 1–4, 6–7; T 2–4, 6–8; Z; Senegal and Mali to Chad and Sudan, Ethiopia and Somalia; Saudi Arabia, Yemen, Oman, Socotra
HAB. Dry bushland, semi-desert scrub, dry woodland and wooded grassland; 0–1200 m
USES. Arrow poison; fish poison; pesticide against cattle lice and ticks. Poisonous to livestock. Cultivated as an ornamental.

SYN. *Nerium obesum* Forssk. in Flora Aegypt.-Arab.: 205 (1775)
 Adenium honghel A.DC. in Prodr. 8: 412 (1844); Stapf in F.T.A. 4, 1: 229 (1902); F.W.T.A. 2: 29 (1963); F.P.U.: 117 (1962). Type: Senegal, *Perrottet* 462 (G-DC, lecto., G, P, isolecto., chosen by Plaizier)
 A. speciosum Fenzl in Sitz. Ber. Acad. Wien, Math. 1, 51: 140 (1865); Stapf in F.T.A. 4, 1: 226 (1902). Type: Ethiopia, near Mt Nubanorum, *Kotschy* 399 (K, holo.)
 A. somalense Balf. f. in Trans. Roy. Soc. Edin. 31: 162 (1888); Stapf in F.T.A. 4, 1: 228, 229 (1902); T.T.C.L.: 47 (1949). Type: Somalia, Coast, *Playfair* 3 (K, holo.)
 A. coetaneum Stapf in F.T.A. 4, 1: 227, 228 (1902); T.T.C.L.: 47 (1949). Type: Sudan?, Bari country, *Speke & Grant* 766 (K, lecto)
 A. tricholepis Chiov., Fl. Somal. 2: 288 (1932). Lectotype: Somalia, Ultra-Juba, *Senni* 70 (FT, lecto., chosen by Plaizier)

NOTE. The shape of the leaves, indumentum and flower size is very variable over the distribution range.

21. **WRIGHTIA**

R. Br. in Mém. Wern. Soc. 1: 73 (1811); Leeuwenberg in Wageningen Agric. Univ. Pap. 87, 5: 35–43 (1987)

Piaggiaea Chiov., Fl. Somala 2: 290 (1932)

Shrubs or trees, occasionally climbers (not in our area); white latex present. Leaves opposite, with colleters in the axils. Inflorescence terminal, dichasial or monochasial, few to many-flowered. Sepals almost free, imbricate, with 1–2 alternate colleters within. Corolla subrotate to infundibuliform; tube cylindrical to campanulate; lobes overlapping to the left; corona variously shaped; stamens exserted or included (not in Africa); anthers narrow, partly fertile, introrse; ovary of 2 carpels, free or slightly connate at the base, united at the apex by the style; pistil head coherent with the anthers, subcapitate or subcylindrical, provided with a basal collar, stigmoid apex minutely bifid. Fruit of two follicles, completely connate (not in our area) or only at

FIG. 24. *WRIGHTIA DEMARTINIANA* — **1, 2**, flowering branches, × ²/₃; **3**, flower, × 2; **4**, opened flower apex, × 8; **5**, pistil, × 8; **6**, fruit, × ²/₃; **7**, open follicle × ²/₃; **8**, seed × ²/₃. 1 from *Cufodontis* 86; 2–5 from *Gillett* 12530; 6 from *Paoli* 680; 7 from *Paoli* 896; 8 from *Paoli* 680. From Meded. Landbouwhogeschool 87(5), drawn by Y.F. Tan, reproduced by permission.

the extreme base and sometimes at the apex, dehiscent throughout by an adaxial slit. Seeds numerous, narrowly fusiform, not beaked, with an apical coma directed towards the base of the fruit; testa thin, smooth; endosperm absent.

An old world genus of 23 species, of which two are indigenous to Africa.

W. demartiniana *Chiov.* in Ann. Bot. Roma 13: 401 (1965); Ngan in Ann. Miss. Bot. Gard. 52: 163 (1965); Leeuwenberg in Wageningen Agric. Univ. Pap. 87, 5: 35, t. 1 (1987); K.T.S.L.: 488, ill., map (1994). Lectotype: Somalia, Juba [Giuba] R. near Matagassile, *Paoli* 839 (FT, lecto., chosen by Ngan)

Shrub or small tree 1.5–5 m high; bark smooth dark grey to black; branchlets pubescent; stipules very short almost reduced to lines. Leaves petiolate; blade elliptic or narrowly elliptic, sometimes obovate, 1.2–8 cm long, 0.3–2 cm wide, rounded or obtuse at the apex, cuneate at the base or decurrent into the petiole, entire, thinly papery when dried, glabrous or puberulous, with obscure reticulate venation when mature; petiole 0.5–2 mm long, pubescent. Inflorescence few-flowered, 1.5–2.5 cm long, pubescent in all parts; peduncle 1–3 mm long; bracts sepal-like; pedicels 4–6 mm long. Flowers very sweetly scented; sepals ovate or broadly ovate, 1.5–2.8 mm long, obtuse, glabrous, with one large squamella inside; corolla tube greenish, 5–8 mm long, almost cylindrical, with stiff recurved hairs inside on the filament ridges; corona white, 1–1.5 mm high, shortly lobed, undulate, glabrous or pubescent outside; lobes white or creamy, oblong, 10–15 mm long, 3–5 mm wide, rounded, spreading; stamens exserted; pistil 4.5–6.5 mm long, glabrous; ovules 30 in each carpel; style rather thick, persistent when corolla is shed, 2–5 mm long. Fruit follicles united at the extreme base, narrowly fusiform, 12–31 cm long, 8 mm wide, pubescent outside, acuminate at the apex, narrowed towards the base; seeds pale brown, 19–25 mm long, 2 mm wide, longitudinally ribbed with coma 3.5–4 cm long, of dirty white smooth hairs. Fig. 24 (p. 70).

KENYA. Northern Frontier District: Samburu Game Reserve, near end of Koitogor, 5 April 1977, *Hooper & Townsend* 1673!; Kitui District: Galana Ranch, Dakadima Hill, 2 May 1975, *Bally* 16697!; Tana River District: 20 km N of Hola, 26 Oct. 1972, *Robertson* 1771
HAB. *Acacia-Commiphora* bushland, often on rocky sites; 50–1200 m
DISTR. **K** 1, 4, 7; Ethiopia, Somalia
USES. Branches used as friction fire-sticks; bark used as scent; latex as laxative

SYN. *Piaggiaea demartiniana* (Chiov.) Chiov., Fl. Somala 2: 291 (1932); K.T.S.: 48 (1961)
    P. boranensis Chiov., Miss. Biol. Borana, Racc. Bot.: 159 (1939). Type: Ethiopia, Borana, Malca Guba sul Daua Parma, *Cufodontis* 86 (FT, holo., W, iso.)
    Wrightia boranensis (Chiov.) Cufod. in B.J.B.B. 30, suppl.: 692 (1960)

22. PLEIOCERAS

Baill. in Bull. Soc. Linn. 1: 759 (1888); Stapf in F.T.A. 4, 1: 165 (1902); Barink in Meded. Landbouwhogeschool 83 (7): 26–42 (1983)

Shrubs, lianas or trees with white latex in bark and leaves. Leaves opposite, petiolate, with colleters in one or two rows in the axils. Inflorescences terminal, paniculate, several- to many-flowered, lax. Flowers 5-merous. Sepals persistent in fruit, connate at the base, ciliate, with colleters within. Corolla salver-shaped, ventricose at the middle; lobes contorted in bud, overlapping to the left, inside with 2 sets of appendages; stamens exserted, covered by appendages; filaments short, hispid towards the apex, with a triangular swelling on the base of the connective, adnate to the pistil head; anthers introrse, sagittate, apiculate, pilose at the apex, partly fertile; pistil glabrous, ovary superior, subglobose, 2-celled, cells free; disk absent; style split at the base; pistil head cylindrical with a thin ring at the base; stigma bifid; ovules numerous. Fruits of 2 spreading almost free, pendulous, follicular

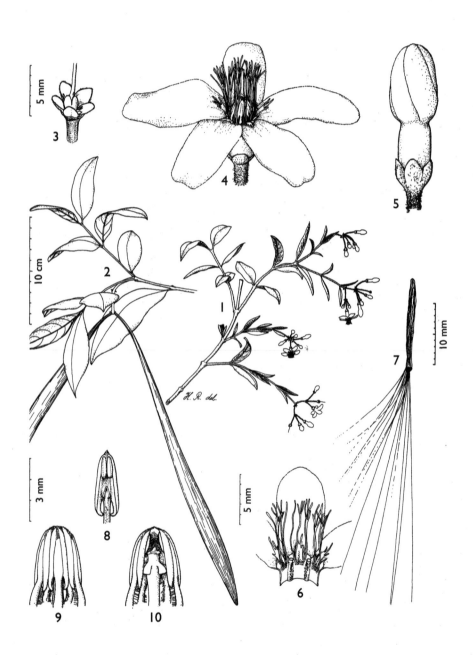

FIG. 25. *PLEIOCERAS ORIENTALE* — **1**, flowering shoot, × ²/₃; **2**, leafy shoot, × 3; **3**, part of calyx and pistil, × 5; **4**, flower, × 2; **5**, bud, × 2; **6**, petals with three sets of appendages, × 3; **7**, seed × ²/₃; **8–10**, anthers and stigma, × 9. 1, 4–5, 8–10 from *Vollesen* 4783; 3 from *Müller & Pope* 1924; 6 from *Vollesen* 4222; 2, 7 from *Simao* 724. From Meded. Landbouwhogeschool 83(7), drawn by H.R., reproduced by permission.

mericarps, slender, dehiscent; pericarp thin, glabrous. Seeds numerous, linear to oblong, with a dense tuft of hairs at the apex directed towards the base of the fruit; embryo large; cotyledons complicate.

Five species in Tropical Africa.

P. orientale *Vollesen* in Bot. Tidsskr. 75: 59 (1980); Barink in Meded. Landbouwhogeschool 83 (7): 37, t. 4 (1983) & in F.Z. 7 (2): 484, t. 114 (1985). Type: Tanzania, Kilwa District, Malemba Thicket, *Vollesen* 4783 (C, holo., BR, EA!, K, LISC, WAG, iso.)

Tree 5–12 m high, deciduous; branchlets densely pilose. Leaves petiolate; blade narrowly ovate or obovate, 3.5–14.5 cm long, 1–5.5 cm wide, puberulous above, tomentose beneath on the main veins; petiole 3–5 mm long, densely pilose. Inflorescence several-flowered, 4.5–6.5 cm long, pubescent on all parts; bracts 2.2–3 mm long, sometimes glabrous; peduncle 1–2 cm long; pedicels 7–15 mm long. Flowers with sepals green, 2–3 mm long, 1–2 mm wide; corolla yellow and violet or dark red, 13.5–20 mm long; tube 6–9 mm long; lobes 7.5–11 mm long, 4–6 mm wide, minutely pubescent outside, papillose inside; corona appendages bright yellow, 7, broomlike and filiform to truncate; stamens inserted 1.5–1.6 mm below the corolla mouth; pistil 8–9 mm long; style 6–8 mm long; ovules 120–150 in each carpel. Fruit dark green, follicles 20–30 cm long, 5–7 mm in diameter; seeds 10–18 mm long, greenish with a dense, 35 mm long tuft of hairs at the apex. Fig. 25 (p. 72).

TANZANIA. Rufiji District: Kichi Hills, 22 Dec. 1976, *Ludanga & Vollesen* 4248!; Kilwa District: Tundu Hills, 28 Dec. 1970, *Ludanga* 1189! & Malemba Thicket, 4 May 1970, *Rodgers* 1051!
DISTR. **K** 7 (see note); **T** 2? (see note), 6, 8; Mozambique
HAB. Deciduous coastal thicket on sand; 300–750 m
USES. None recorded

NOTE. *Bally* 11537 from **T** 2, Masai District: between Endulen and Lake Eyassi at 1700 m is possibly this species. The material consists of follicles only.
   Luke (pers. comm.) has seen this species in **K** 7, Kwale District: Mwele Forest, in Oct. 2001. The material was glabrous, similar to that of *Pedro & Pedrogao* 5225 from Niassa, Mozambique.

23. STEPHANOSTEMA

K. Schum. in E.J. 34: 325 (1904); Barink in Meded. Landbouwhogeschool 83 (7): 42–47 (1983)

Shrub with white latex. Leaves opposite, with a row of colleters in the axils. Inflorescence terminal or in axils of upper leaves, paniculate. Flowers 5-merous. Calyx with 0–5 colleters at the base near the edge. Corolla salver-shaped; tube urceolate; lobes overlapping to the left, with a corona at the base; stamens included; filaments short, hispid at the apex; anthers apiculate and pilose at the apex, rounded at the base; ovary of two free, rounded carpels, glabrous; pistil head with a cylindrical basal part with a skirt at the top and bottom and a bifid stigmoid apex. Fruit of two slender, almost free follicular mericarps, dehiscent adaxially, glabrous. Seeds with a coma at the apex, directed towards the base of the fruit

Monotypic.

S. stenocarpum *K. Schum.* in E.J. 34: 325 (1904); T.T.C.L.: 55 (1949); Barink in Meded. Landbouwhogeschool 83 (7): 42, t. 6, phot. 2–5 (1983). Types: Tanzania, Uzaramo District, Sachsenwald, near Dar es Salaam, *Engler* 2156 (B†, holo.); neotype: Dar es Salaam, Gongolamboto, *Wingfield* 4224 (K!, neo.; DSM, EA, MO, isoneo., chosen by Barink)

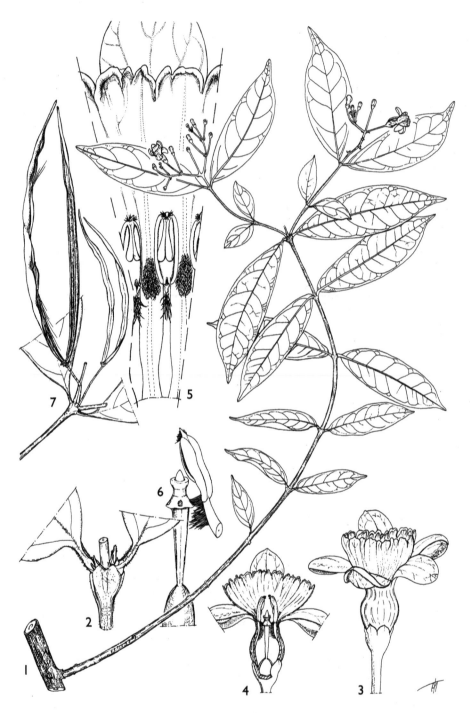

FIG. 26. *STEPHANOSTEMA STENOCARPUM* — **1**, flowering branch, × ²/₃; **2**, branches with colleters, × 4; **3**, flower, × 4; **4**, opened flower, × 10; **5**, part of corolla, × 10; **6**, pistil and part of stamen, × 10; **7**, fruit, × ²/₃. 1,7 from *Wingfield* 4224; 2 from *Harris* 3071; 3–6 from *Wingfield* 3616. From Meded. Landbouwhogeschool 83(7), drawn by Y.F. Tan, reproduced by permission.

Erect shrub 0.25–2 m high, with white latex; bark rough, dark brown with pale lenticels, glabrous, branchlets red to brown, terete, often sulcate when dry, sparsely and minutely pubescent. Leaves petiolate; blade ovate, obovate or elliptic, 4.8–9.5 cm long, 1.7–3.7 cm wide, acuminate at the apex, rounded or cuneate at the base, glabrous; petiole 2–3 mm long. Inflorescence lax, 1.5–2 cm long, glabrous on all parts; bracts up to 2 mm long, acute, with colleters within; peduncle 2.5–3 mm long; pedicels 5–9 mm long. Flowers with sepals ovate, 0.8–1.5 mm long, acute, glabrous; corolla tube yellow, 3–4 mm long; lobes 4.2–5 mm long, 2.5–2.8 mm wide, rounded; corona white, 3 mm long, undulate; pistil 4–4.5 mm long; style filiform, 2.2 mm long, glabrous; ovules about 18 in each carpel. Fruit pale green to pale grey, 10–15 cm long, 3–5 mm wide; seeds 9–10 mm long; coma 18–20 mm. Fig. 26 (p. 74).

TANZANIA. Uzaramo District: Dar es Salaam Airport, 10 Aug. 1969, *Harris* 3071! & Gongolamboto Graveyard on Kisarawe road near the Airport, 1 Jan. 1986, *Lovett* 500! & 10 Aug. 1993, *Luke et al.* 3754!
DISTR. **T** 6; not known elsewhere
HAB. Relic wooded patch; 0–60 m
USES. None recorded

NOTE. Cultivated at the Royal Botanic Gardens, Kew and at Wageningen Botanic Gardens.

24. STROPHANTHUS*

DC. in Bull. Soc. Philom. 64: 122 (1802); Gilg, Monogr. Afr. Pfl.-Fam. & Gatt. 7: 7 (1903); Beentje in Meded. Landbouwhogeschool 82 (4): 17 (1982)

Zygonerion Baill. in Bull. Soc. Linn. Paris 1: 758 (1888)

Shrubs, sarmentose shrubs, or lianas, rarely trees (but not in our area); latex present in most parts; bark rough, sometimes with corky protuberances, lenticellate; stipules reduced to small rims. Leaves opposite or ternate, rarely quaternate, with colleters in the axil of the petiole. Inflorescence terminal or in forks of branches, rarely axillary, 1–many-flowered in simple or compound dichasial cymes; bracts sepal-like or scarious. Flowers 5-merous. Calyx imbricate-quincuncial, deciduous in fruit, colleters present adaxially. Corolla often with long tails to the lobes, the tube cylindrical proximally and slightly wider distally becoming infundibuliform, at the mouth with a 10-lobed corona, the lobes erect and in pairs at the margins of the corolla lobes, lobes alternisepalous, overlapping to the right, ovate to oblong, at the apex acute to tailed; stamens included or partly exserted, connate to the style and forming a cone around the distal part of the style and stigma; ovary 2–celled, the cells partly connate, style terete, stigma subtended by an angular clavuncula. Fruit of 2 woody mericarps on a thickened woody pedicel, the mericarps usually divergent. Seeds many per mericarp, narrowly ovoid with a beaked apical coma.

Thirty-eight species in Africa, Madagascar and tropical Asia.

Key: note that several *Strophanthus* species may flower before the leaves appear: *S. eminii, S. petersianus, S. sarmentosus.* The last species only occurs in **U** 2, while the first two do not – these can be distinguished as follows: *S. eminii* is found above 600 m altitude and has puberulous branchlets, while *S. petersianus* occurs below 400 m and has glabrous branchlets.

1. Branchlets and leaves hispid · 2
   Branchlets and leaves glabrous, puberulous or (*S. hypoleucos*) tomentellous or (*S. preussii*) rarely scabrid · · · · · · · · · · · · · 3

* by H. J. Beentje

2. Sepals unequal, the outer much wider than the inner;
   western species (**U** 2, 4; **T** 4) · · · · · · · · · · · · · · · · ·   10. *S. hispidus*
   Sepals subequal; eastern species (**K** 7; **T** 3, 6, 8) · · · · · ·   11. *S. kombe*
3. Leaves with dense white tomentum beneath; branchlets
   densely puberulous to tomentellous · 4
   Leaves glabrous or nearly so; branches glabrous or nearly
   so, or (in *S. mirabilis*) densely puberulous · 5
4. Corolla lobe tails 9–17 cm long; fruit with shaggy
   protuberances ·   5. *S. eminii*
   Corolla lobe tails 1.5–5 cm long; fruit smooth, without
   protuberances ·   7. *S. hypoleucos*
5. Branchlets densely puberulous; densely branched small
   shrub of dry bushland in Eastern Kenya · · · · · · · · · ·   6. *S. mirabilis*
   Branchlets glabrous or nearly so; sarmentose shrubs or
   lianas, rarely a small tree (*S. welwitschii*) · 6
6. Flowers present · 7
   Fruits present · 13
FLOWERS
7. Corolla lobes without tails · 8
   Corolla lobes with filiform tails · 9
8. Branches with corky ridges; corona lobes 2–6 mm high   1. *S. courmontii*
   Branches smooth; corona lobes 5–23 mm high · · · · · ·   2. *S. welwitschii*
9. Branches with corky ridges; corona lobes > (5–)7 mm · · · · · · · · · · · · · · 10
   Branches smooth; corona lobes < 6.5 mm · 11
10. Ovary and inside of corolla lobes minutely puberulous;
    coastal species below 400 m · · · · · · · · · · · · · · · · · ·   4. *S. petersianus*
    Ovary and corolla lobes glabrous; western species above
    700 m; **U** 2 ·   3. *S. sarmentosus*
11. Tails of corolla 12–18 cm long; western species above
    1200 m ·   8. *S. preussii*
    Tails of corolla 4.5–7 cm long · 12
12. Sepals 14–21 mm long; eastern species below 800 m · · ·   9. *S. zimmermannii*
    Sepals 5–12 mm long; western species above 1400 m · ·   12. *S. bequaertii*
FRUITS
13. Fruit mericarps with an obtuse apex · 14
    Fruit mericarps tapering but at the apex widening into a
    small knob · 17
14. Branches smooth; branchlets minutely puberulous;
    western Tanzania (**T** 4) ·   2. *S. welwitschii*
    Branches with corky protuberances; branchlets usually
    glabrous · 15
    (branches smooth and glabrous rare specimens of 12. *S. bequaertii*)
15. (NOTE: these three species are almost impossible to distinguish in fruit)
    Western species, found in Uganda (**U** 2) · · · · · · · · · · ·   3. *S. sarmentosus*
    Eastern species, found along Kenya and Tanzania coasts · · · · · · · · · · · · · 16
16. Mericarps 12–26 cm long, 3–4.5 cm in diameter · · · · · ·   1. *S. courmontii*
    Mericarps 20–37 cm long, 2–3.5 cm in diameter · · · · · ·   4. *S. petersianus*
17. Branches with corky protuberances; seed beak not
    bearing coma for more than 18 mm; coastal species,
    found below 500 m ·   4. *S. petersianus*
    Branches smooth; seed beak not bearing coma for less than
    15 mm (except *S. welwitschii* from **T** 4 and *S. bequaertii*
    from **U** 2) · 18
18. Mericarps 32–40 cm long, ± 1.2 cm in diameter; coastal
    species from Kenya and Tanzania, found below 800 m ·   9. *S. zimmermannii*
    Mericarps 10–34 cm long, 1–3 cm in diameter; western
    species found above 1200 m · 19

19. Branchlets glabrous, rarely scabrid · 20
    Branchlets minutely puberulous; seed beak not bearing
    coma for 17–54 mm ·   2. *S. welwitschii*
20. Seed beak not bearing coma for 3–8 mm · · · · · · · · · · ·   8. *S. preussii*
    Seed beak not bearing coma for 15–20 mm · · · · · · · · ·   12. *S. bequaertii*

The following species occur just outside our area, and might be found within our area: *S. gardeniiflorus* (a liana with flowers without tails, from gallery forest in S Congo (Kinshasa) and N Zambia), *S. holosericeus* (a liana resembling *S. eminii* but with very small protuberances on the fruit, from forest near waterfalls in SE Congo (Kinshasa) and N Zambia), and *S. nicholsonii* (a shrub or small liana with short-tomentose leaves and tailed flowers from miombo woodland in N Malawi and N Mozambique)

The following species have been cultivated in East Africa:
S. amboensis (Schinz) Engl. & Pax, in Kampala (Feb. 1929, *Liebenberg* 762!). A glabrous liana from Angola; corolla orange-yellow turning dark purple-red, streaked white inside.
S. caudatus (L.) Kurz in Amani (T.T.C.L.: 55 (1949); Jan. 1950, *Verdcourt* 33!). Asian liana with long tails to the corolla, which is white turning yellow to red, the tails turning purple-red.
S. gratus (Wall. & Hook.) Baill. in Dar es Salaam (State House grounds, Aug. 1972, *Ruffo* 503!), Amani (T.T.C.L.: 56 (1949) and Nairobi (Jex-Blake's garden, Jan. 1954, *Bally* 9454!). A glabrous liana from W Africa, corolla without tails, white turning yellow to red-purple.

1. **S. courmontii** *Franch.* in Journ. de Bot. 7: 300 (1893); Stapf in F.T.A. 4, 1: 182 (1902); T.T.C.L.: 55 (1949); Beentje in Meded. Landbouwhogeschool 82 (4): 62, t. 14 (1982) & in F.Z. 7 (2): 470, t. 111 (1985); K.T.S.L.: 485, map (1994). Type: Tanzania, Nguvu (?Nguru) Mts, *Sacleux* 2032 (P!, holo.)

Liana, 5–22 m long, or less often a sarmentose shrub 0.6–4 m high, deciduous, with the flowers appearing after the leaves; trunk to 10 cm in diameter, with corky ridges to 1.8 cm high and 5 cm long; branches with compressed corky triangles at the nodes; branchlets glabrous; latex white. Leaves opposite; blade elliptic or ovate to obovate, 2–14 cm long, 2.5–6.5 cm wide, base rounded or cuneate, apex mucronate or acuminate, glabrous; petiole 3–11 mm long. Inflorescence sessile or pedunculate, 1–3-flowered, glabrous or sparsely puberulous in all parts; pedicels 1–7.5 mm long. Flowers fragrant; sepals ovate, 3–10 mm long, acute or apiculate; corolla tube white and red, 22–43 mm long, corolla lobes ovate, 20–57 mm long, 10–27 mm wide, acuminate, corona lobes 2–6 mm long. Fruit grey- or purplish black, mericarps ± opposite-divergent, narrowly ellipsoid, 12–26 cm long, 3–4.5 cm in diameter, apex obtuse, glabrous; seed 10–15 mm long, densely pubescent, with a stalked coma 5–12 cm long.

Kenya. Tana R. District: Congolani, 12 Mar. 1990, *Luke et al. TPR* 136!; Kilifi District: Rabai, *Joanna* in EA 10528!; Kwale District: 6 km from Ramisi R. on Mrima Hill road, 23 Mar. 1974, *Faden* 74/301!
Tanzania. Lushoto District: Fanusi, 25 Nov. 1941, *Greenway* 6676!; Morogoro District: Turiani, Nov. 1953, *Semsei* 1439!; Iringa District: Lukosi R., 22 Oct. 1930, *Burtt* 6071!
Distr. **K** 7; **T** 3, 6–8; Zambia, Malawi, Zimbabwe, Mozambique
Hab. Gallery forest, riverine thickets, coastal forest; 0–1400 m
Uses. Used for walking sticks

Syn. *S. courmontii* Franch. var. *fallax* Holmes in Pharm. J.4 (12): 487 (1901). Type: Malawi, *Buchanan* 1219 (K!, holo.; BM!, E, iso.)
   S. courmontii Franch. var. *kirkii* Holmes in Pharm. J.4 (12): 487 (1901). Type: Tanzania, District unknown, Yao forest, *Bishop Steere* s.n. (K!, holo.)

Note. Also cultivated in Kampala and Nairobi.

2. **S. welwitschii** (*Baill.*) *K. Schum.* in E.P. 4 (2): 59 (1900); Gilg, Monogr. Afr. Pfl.-Fam. & Gatt. 7: 21, t. 8 (1903); Beentje in Meded. Landbouwhogeschool 82 (4): 154, t. 43 (1982) & in F.Z. 7 (2): 479 (1985). Type: Angola, between Pungo Andongo and Cuanza R., *Welwitsch* 5991 (P!, holo., BM!, BR!, C, G, K!, LE, LISU, MO, P!, iso.))

Sarmentose shrub or small tree, 0.6–5 m high, or liana to 8 m long, deciduous; trunk to 10(–40?) cm diameter, bark dark brown or grey; branchlets minutely puberulous; latex white or clear. Leaves opposite, or in some branches ternate to quaternate; blade ovate, narrowly elliptic or rarely slightly obovate, to 8.5 cm long, to 4 cm wide, base cuneate or rounded, margins slightly revolute, apex rounded, acute or acuminate, glabrous or sparsely puberulous; petiole 1–5 mm long. Inflorescence 1–5-flowered, sessile or pedunculate, glabrous or puberulous in its parts; pedicels 3–17 mm long. Flowers fragrant; sepals (narrowly) ovate, 5–19 mm long, acute; corolla tube white to pink proximally, reddish purple distally, creamy and streaked with red inside, 13–38 mm long, corolla lobes white to pale pink but much darker on one side, narrowly triangular, 10–48 mm long, 8–29 mm wide, acute, not tailed, corona lobes 5–23 mm long. Fruit dark- or purple-brown, hard, the mericarps ± opposite-divergent, narrowly ellipsoid, 10–34 cm long, 1–2.5 cm in diameter, tapering and ending in an obtuse point or small knob, lenticellate, glabrous; seeds 8–20 mm long, densely pubescent, with a stalked coma 5–15 cm long.

TANZANIA. Ufipa District: Tatanda mission, 24 Feb. 1994, *Bidgood et al.* 2437! & Nov. 1994, *Goyder et al.* 3745! & Ngorotwa, sine die, *Carmichael* 1001!
DISTR. **T** 4; Congo (Kinshasa), Angola, Zambia
HAB. Miombo woodland and gallery forest, often in rocky sites; 1200–1800 m
USES. Source of arrow poison in neighbouring countries

SYN. *Zygonerion welwitschii* Baill. in Bull. Mens. Soc. Linn. Paris 1: 758 (1888)
Strophanthus verdickii De Wild., Etudes Fl. Katanga: 103, t. 21 (1903). Type: Congo (Kinshasa), Shaba, Lukafu, *Verdick* 84 (BR!, holo.)
S. verdickii De Wild. var. *latisepalus* De Wild., Etudes Fl. Katanga: 104 (1903). Type: Congo (Kinshasa), Shaba, Lukafu, *Verdick* 146 (BR!, holo.)
S. gilletii De Wild., Etudes Fl. Katanga: 105, t. 21 (1903). Type: Congo (Kinshasa), lower Congo R., Kimuenza, *Gillet* 2129 (BR!, holo.)
S. katangensis Staner in Ann. Soc. Sci. Brux. ser. B 52: 94 (1932). Type: Congo (Kinshasa), Shaba, Dilolo, *De Witte* 607 (BR!, holo., BR!, NY, iso.)

3. **S. sarmentosus** *DC.* in Bull. Soc. Philom. Paris 64: 123, t. 1 (1802); Stapf in F.T.A. 4, 1: 180 (1902); Huber in F.W.T.A. ed. 2, 2: 70, t. 215 (1963); Beentje in Meded. Landbouwhogeschool 82 (4): 133, t. 36 (1982). Type: Sierra Leone, sine loc., *Smeathman* s.n. (G-DC, holo., BM!, FI-W, P-JU!, UPS, iso.)

Sarmentose shrub 0.5–4 m high, or liana to 40 m long, deciduous, the flowers appearing before or with the leaves; trunk to 17 cm in diameter, bark pale brown, corky, deeply fissured; branches lenticellate, with many corky protuberances to 1 cm high; branchlets dark or reddish brown, glabrous or rarely minutely puberulous. Leaves opposite, ternate, or rarely quaternate; blade elliptic or ovate, 2–15 cm long, 1.5–7 cm wide, base rounded or cuneate, apex acuminate, glabrous; petiole 2–21 mm long. Inflorescence terminal or in forks, 1–11-flowered, sessile or pedunculate, congested; pedicels 2–12 mm long. Flowers fragrant; sepals purple and green, ovate or elliptic, 5–20 mm long, acute; corolla tube white and turning yellow proximally, pink and turning purple distally, the inside white streaked with red, 17–40 mm long, corolla lobes white turning yellow, ovate, 7–20 mm long, 6–18 mm wide, narrowing into the pale yellow 4–12 cm long pendulous tails, corona lobes 5–22 mm long. Fruit brown or purple-brown, the mericarps ± opposite-divergent, narrowly ovoid, 10–28 cm long, 1.4–4.4 cm in diameter, slightly sulcate, lenticellate, glabrous; seeds 8–20 mm long, densely pubescent, with a stalked coma 4–16 cm long.

var. **sarmentosus**

Inflorescence minutely puberulous in all parts. Sepals 6.5–20 mm long. Fruit diameter 2.2–4 cm, apex obtuse; seed beak glabrous for 10–62 mm.

UGANDA. Toro District: Bwamba forest, 1 Jan. 1945, *Greenway & Eggeling* 7061! & Kionsozi forest, sine die, *Dawe* 20! & Mawokota, Feb. 1905, *E. Brown* 162!
DISTR. U 2; W and central Africa, from Senegal and Mali to Central African Republic and Congo (Kinshasa)
HAB. Moist forest and gallery forest; 750–1300 m
USES. Source of arrow poison in W and central Africa

SYN. *S. sarmentosus* DC. var. *major* Dewèvre in J. Pharm. Anvers 50: 428 (1894). Type: Congo (Kinshasa), lower Congo R., Congo di Vanga, *Laurent* s.n. (BR, holo., lost); neotype: Congo (Kinshasa), upper Congo R., Yangambi, Isalowe Forest Reserve, *Louis* 13648 (BR, neo., B, K!, LISC!, P!, isoneo., chosen by Beentje)
  S. sarmentosus DC. var. *pubescens* Staner & Michotte in B.J.B.B. 13: 52 (1934), except for leaves. Type: Congo (Kinshasa), upper Zaire, Avakubi, *Bequaert* 2062 (BR!, holo.)

NOTE. The second variety, var. *glabriflorus* Monach., is resticted to a small area in W Africa.

4. **S. petersianus** *Klotzsch* in Reise Mossamb. Bot. 1: 276 (1861); Stapf in F.T.A. 4, 1: 182 (1902); T.T.C.L.: 58 (1949); Codd in F.S.A. 26: 291 (1963); Beentje in Meded. Landbouwhogeschool 82 (4): 120, t. 33 (1982) & in F.Z. 7 (2): 477 (1985); K.T.S.L.: 486 (1994). Type: Mozambique, Tete, Zambezi R., *Peters* s.n. (B†, holo., K!, iso.)

Sarmentose shrub or liana, 1–15 m long, deciduous, flowers appearing with or rarely before the leaves; trunk to 10 cm in diameter, bark pale grey; branches with corky flattened triangular protuberances to 2.5 cm high at the nodes or rarely in between; branches glabrous or rarely puberulous; latex white or reddish. Leaves petiolate, blade elliptic or ovate, 3–11 cm long, 1.7–5 cm wide, base cuneate or rounded, apex acuminate, glabrous or very rarely sparsely puberulous; petiole 2–13 mm long. Inflorescence 1–4-flowered, sessile or pedunculate, glabrous or occasionally puberulous in its parts; pedicels 3–11 mm long. Flowers fragrant, white turning yellow, near the mouth outside and tails outside red; sepals unequal, ovate or narrowly elliptic, 5–21 mm long, acute; corolla tube 13–37 mm long, corolla lobes ovate, 9–16 mm long, 6–15 mm wide, narrowing into the 8–20 cm long pendulous tails, corona lobes 6–15 mm long. Fruit dark brown, hard, the mericarps ± opposite-divergent, narrowly ovoid, 20–37 cm long, 2–3.5 cm in diameter, tapering to an obtuse point or small knob, lenticellate, glabrous; seeds 10–18 mm long, densely pubescent, with a stalked coma 6–16 cm long.

KENYA. Kilifi District: Arabuko, 1929, *Graham* 1712! & Ganze, 28 Dec. 1985, *Robertson* 4122!; Kwale District: Muhaka, 14 Apr. 1977, *Gillett* 21064!
TANZANIA. Pangani District: Msubugwe Forest Reserve, 3 Mar. 1963, *Mgaza* 552!; Uzaramo District: Mogo Forest Reserve, 11 Oct. 1965, *Mgaza* 722!; Rufiji District: Mafia I., 2 Oct. 1937, *Greenway* 5370!
DISTR. K 7; T 3, 6, 8; Zambia, Malawi, Mozambique, Zimbabwe, South Africa
HAB. Coastal forest and woodland, often in rocky sites; 0–650 m
USES. Source of arrow poison

SYN. *S. petersianus* Klotzsch var. *grandiflorus* N. E. Br. in K.B. 1892: 126 (1892); Gilg in Monogr. Afr. Pfl.-Fam. & Gatt. 7: 28 (1903). Type: Mozambique, Delagoa Bay, *Monteiro* 1 (K!, holo., FT, G, P!, W!, iso.)
  S. sarmentosus DC. var. *verrucosus* Pax in E.J. 15: 374 (1892). Type: Kenya, coast near Mombasa, *Hildebrandt* 1976 (B†, holo., BM!, K!, L!, LE, NY, P!, W!, WU, iso.)
  S. grandiflorus (N. E. Br.) Gilg in E.J. 32: 161, t. 7 (1902); Stapf in K.B. 1907: 510 (1907)
  S. verrucosus (Pax) Stapf in F.T.A. 4, 1: 181 (1902); T.T.C.L.: 56 (1949)

5. **S. eminii** *Aschers. & Pax* in E.J. 15: 366, t. 10–11 (1892); Stapf in F.T.A. 4, 1: 172 (1902); Gilg in Monogr. Afr. Pfl.-Fam. & Gatt. 7: 39 (1903); T.T.C.L.: 55 (1949); Verdcourt & Trump, Comm. Poison. Pl. E. Afr.: 137, t. 10/a–d (1969); Beentje in Meded. Landbouwhogeschool 82 (4): 69, t. 16 (1982) & in F.Z. 7 (2): 472 (1985). Type: Tanzania, Kondoa District, Irangi, *Fischer* 382! (B, holo.†, K!, iso.)

Shrub or small tree 1–7 m high, sometimes with branches climbing to 10 m high, deciduous, the flowers appearing before or with the leaves; trunk to 6 cm in diameter; branches slightly fleshy, smooth or sulcate, densely puberulous; latex clear, white or yellow. Leaves opposite, petiolate, blade silvery abaxially, (broadly) ovate or elliptic, 6–24 cm long, 4–18 cm wide, base cuneate or rounded, rarely subcordate, apex acute or acuminate, pubescent and glabrescent above, tomentose beneath; petiole 1–10 mm long. Inflorescence on leafless shoots, apparently axillary, rarely in forks, 1–12-flowered, congested, densely pubescent in all parts; pedicels 1–8 mm long. Flowers fragrant, white turning yellow and then red; sepals ovate, 11–25 mm long, acute; corolla tube 17–26 mm long, corolla lobes ovate, 7–15 mm long, 4.5–10 mm wide, narrowing gradually into filiform tails 9–17 cm long, corona lobes 2–7 mm long. Fruit pale brown, mericarps ± opposite-divergent, narrowly ellipsoid, 20–38 cm long, 1.5–3.2 cm in diameter, tapering to an obtuse apex or a knob, densely shaggy with villous protuberances 4–18 mm long; seeds 11–24 mm long, densely pubescent, with a stalked coma 8–17 cm long. Fig. 27 (p. 81).

TANZANIA. Mwanza District: Mbarika, 20 Mar. 1952, *Tanner* 590!; Kondoa District: Turu, 18 Feb. 1982, *Sigara* 243!; Iringa/Kilosa District: Ruaha Gorge, Apr. 1966, *Procter* 3305!
DISTR. T 1, 4–7; Congo (Kinshasa), Zambia
HAB. Miombo woodland or *Acacia-Commiphora* bush, especially in rocky places; 500–1500 m
USES. Roots medicinal (emetic); leaves as toilet paper; seeds and roots source of arrow poison

SYN. *S. wittei* Staner in Ann. Soc. Sci. Brux. ser. B, 52: 90 (1932). Type: Congo (Kinshasa), Shaba, Kiamba, *De Witte* 280 (BR!, holo., BR!, NY, iso.)
    S. eminii Asch. & Pax var. *wittei* (Staner) Staner & Michotte in B.J.B.B. 13: 34 (1934)

6. **S. mirabilis** *Gilg* in E.J. 32: 32 (1902); Stapf in F.T.A. 4, 1: 186 (1902); Gilg in Monogr. Afr. Pfl.-Fam. & Gatt. 7: 27, t. 6 (1903); Beentje in Meded. Landbouwhogeschool 82 (4): 106, t. 28 (1982); K.T.S.L.: 485, ill., map (1994). Type: Kenya, Northern Frontier District, Gave Libin near Wonte, *Ellenbeck* 2205 (B†, holo., K!, lecto., chosen by Beentje)

Shrub 1–3 m high, densely branched, occasionally with lianescent branches to 4.5 m high; deciduous, the flowers appearing before or with the leaves; branches brown, lenticellate; branchlets densely puberulous; latex amber. Leaves opposite, subsessile or with a petiole to 3 mm long, blade narrowly elliptic, 0.8–3.6 cm long, 0.3–1.4 cm wide, base cuneate, margins undulate, apex obtuse or rounded, glabrous but for the hispidulous midrib and margins. Inflorescence on branches or on short-shoots, 1(–3)-flowered, all parts densely puberulous; pedicels 1–8 mm long. Flowers with sepals erect or spreading, elliptic to narrowly obovate, 6–15 mm long; corolla white suffused with pink and turning yellow, the tube 6–13 mm long, corolla lobes ovate, 4–8 mm long, 3–6 mm wide, narrowing into the 4–8 cm long tails, corona lobes yellow and red or chocolate brown-spotted, 2.5–6.5 mm long. Fruit brown to purplish brown, the mericarps opposite-divergent, ± cylindrical, 16–32 cm long, 1–2 cm in diameter, tapering and ending in a small knob, sulcate, lenticellate, glabrous; seeds 9–21 mm long, densely pubescent, with a stalked coma 7–18 cm long.

KENYA. Northern Frontier District: 80 km SW of Mandera on El Wak road, 26 May 1952, *Gillett* 13322! & Wajir, 22 June 1951, *Kirrika* 65!; Kitui District: 2 km on Galana Ranch–Voi road, *Agnew et al.* 7348!
DISTR. K 1, 4, 7; Somalia
HAB. *Acacia-Commiphora* bushland; 0–600 m
USES. Roots edible when cooked?

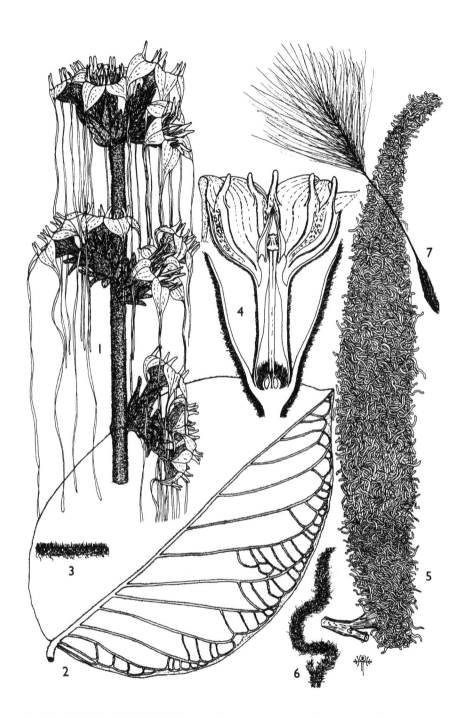

FIG. 27. *STROPHANTHUS EMINII* — **1**, flowering branch, × ²⁄₃; **2**, leaf × ²⁄₃; **3**, leaf cross-section, detail, × 6; **4**, section of flower, × 2; **5**, fruit with one follicle removed, × ²⁄₃; **6**, follicle protuberance, × 6; **7**, seed × ²⁄₃. 1 from *Leach* 10086; 2–3 from *Peter* 34769; 4 from *Mhoro* 1150; 5–6 from *De Witte* 5795; 7 from *Robijns* 1961. From Meded. Landbouwhogeschool 82(4), drawn by H. Beentje.

7. **S. hypoleucos** *Stapf* in K.B. 1914: 81 (1914); T.T.C.L.: 56 (1949), as *hypoleucus*; Beentje in Meded. Landbouwhogeschool 82 (4): 93, t. 23 (1982) & in F.Z. 7 (2): 474 (1985). Type: Mozambique, Mt M'kota, *Stocks* 148 (K!, holo. & iso.)

Shrub 1–4 m high, rarely with lianescent branches; deciduous, the flowers appearing with the leaves; branchlets densely tomentellous; latex white. Leaves opposite, petiolate, blade circular, elliptic or ovate to obovate, 2–8 cm long, 1.3–8.5 cm wide, base cuneate, rounded or subcordate, apex acute, rounded or slightly emarginate, densely short-pubescent above, tomentose with whitish hairs beneath; petiole 2–8 mm long. Inflorescence 1–6-flowered, pedunculate, congested, densely tomentellous in all parts; pedicels 4–15 mm long. Flowers with sepals elliptic to obovate, 5–13 mm long, acute; corolla tube 14–22 mm long, corolla lobes ovate, 6–12 mm long, 5–11 mm wide, narrowed into the 1.5–5 cm long filiform tails, corona lobes 0.6–2 mm long. Fruit chocolate brown, the mericarps ± opposite-divergent, cylindrical, 12–23 cm long, ± 2 cm in diameter, hard, lenticellate, pubescent but glabrescent; seeds 8–13 mm long, densely pubescent, with a stalked coma 4–8 cm long.

TANZANIA. Masasi District: Pangani hill, 11 Mar. 1991, *Bidgood et al.* 1904! & 75 km from Tunduru to Masasi, 19 Nov. 1966, *Gillett* 17920! & Nangua, Nov. 1951, *Eggeling* 6388!
DISTR. **T** 8; N Mozambique
HAB. On rocks in woodland; 450–550 m
USES. None recorded

8. **S. preussii** *Engl. & Pax* in E.J. 15: 369 (1892); Stapf in F.T.A. 4, 1: 176 (1902); Gilg in Monogr. Afr. Pfl.-Fam. & Gatt. 7: 24, t. 4 (1903); Huber in F.W.T.A. ed. 2, 2: 70 (1963); Beentje in Meded. Landbouwhogeschool 82 (4): 125, t. 34 (1982). Type: Cameroon, Barombi ravine, *Preuss* 116 (B†, holo., K!, lecto., HBG, M!, PRE, iso., chosen by Beentje)

Sarmentose shrub 0.7–5 m high, or more often a liana, 1–12 m long, ?evergreen; trunk to 2.5 cm in diameter; branches reddish or purple-brown, lenticellate; branchlets brown, glabrous or rarely scabrid; latex clear or white. Leaves opposite, petiolate, blade elliptic or ovate to obovate, 2–19 cm long, 1.5–7.5 cm wide, base cuneate, rounded or rarely subcordate, apex acuminate, glabrous or rarely scabrid; petiole 2–14 mm long. Inflorescence terminal or in forks, 1–48-flowered, sessile or pedunculate, lax, minutely puberulous or rarely glabrous in all parts; pedicels 4–25 mm long. Flowers fragrant, white, turning yellow and then orange, suffused with red near the mouth, spotted and streaked with red inside; sepals unequal, the outer ovate, the inner linear, 4–25 mm long, acute or obtuse; corolla tube 12–26 mm long, corolla lobes ovate, 4–12 mm long, 3.5–10 mm wide, narrowing into the 12–18 cm long pendulous tails, corona lobes 1–2.5 mm long. Fruit dark brown, hard, the mericarps ± opposite-divergent, ± cylindrical, 15–29 cm long, 1–3 cm in diameter, tapering and ending in a knob, sulcate, lenticellate, glabrous; seeds 12–20 mm long, densely puberulous to pubescent, with a shortly beaked coma 6–10 cm long. Fig. 28 (p. 83).

UGANDA. Bunyoro District: Budongo Forest, Mar. 1933, *Eggeling* 1282!; Toro District: Bwamba Forest, 2 Feb. 1945, *Greenway & Eggeling* 7067!; Masaka District: Malabigambo Forest, 2 Oct. 1953, *Drummond & Hemsley* 4580!
TANZANIA. Bukoba District: Bugandika, 1935, *Gillman* 438!; Mingiso Forest Reserve, *Watkins* 521! & Minziro, 24 Apr. 1994, *Congdon* 358!
DISTR. **U** 1, 2, 4; **T** 1; West Africa from Guinea and Sierra Leone to Central African Republic, Congo (Kinshasa) and Angola
HAB. Moist forest, swamp forest and forest margins; 750–1200 m
USES. Source of arrow poison in Congo; stem fibres for fishing nets

SYN. *S. preussii* Engl. & Pax var. *brevifolius* De Wild., Not. Pl. Utiles Congo: 248 (1908). Type: Congo (Kinshasa), Bandundu, Bena Dibele, *Flamigni* 190 (BR!, holo.)
    S. preussii Engl. & Pax var. *scabridulus* Monach. in Phytologia 3 (9): 478 (1951). Type: Congo (Kinshasa), upper Congo, Epulu area, *Putman* 118 (A, holo., BR!, iso.)

FIG. 28. *STROPHANTHUS PREUSSII* — **1**, flowering branch × ²/₃; **2**, leaf × ²/₃; **3**, section of flower, × 2; **4**, outermost petal, × 2; **5**, innermost petal , × 2; **6**, side view of stamen, × 4; **7**, fruit, × ²/₃; **8**, seed × ²/₃. 1 from *Leeuwenberg* 2892; 2–6 from *Beentje* 182; 7 from *Gerard* 2698; 8 from *Brass & Woodward* 20951. From Meded. Landbouwhogeschool 82(4), drawn by J. Williamson & H. Beentje.

9. **S. zimmermannii** *Monach.* in Phytologia 3 (9): 477 (1951); Beentje in Meded. Landbouwhogeschool 82 (4): 161, t. 45 (1982); K.T.S.L.: 486 (1994), as *zimmermannianus.* Type: Tanzania, Lushoto District, Gonja [Ngonya] Mt, *Zimmermann* 1496 (EA!, holo., K!, NY, iso.)

Shrub or liana to 6 m long; branches lenticellate; branchlets glabrous; latex white. Leaves opposite; blade ovate or elliptic, 8–18 cm long, 3–7.5 cm wide, base rounded or subcordate, apex acuminate, glabrous; petiole 3–6 mm long. Inflorescence terminal or in forks, 1–7-flowered, sessile or pedunculate, lax, puberulous in all parts; pedicels 7–12 mm long. Flowers with sepals unequal, the outer wider and shorter than the inner, 14–21 mm long; corolla white turning yellow; corolla tube 18–26 mm long, corolla lobes ovate, 6–8 mm long, 4–7 mm wide, narrowing into the 4.5–5.5 cm long pendulous tails, corona lobes 2.5–4.5 m long. Fruit reddish brown, the mericarps divergent, ± cylindrical, 32–40 cm long, ± 1.2 cm in diameter, long-tapering and ending in a small knob, lenticellate, puberulous but glabrescent; seeds 10–15 mm long, densely puberulus, with a stalked coma 5–9 cm long.

Kenya. Kwale District: Pengo Hills, 27 Mar. 1968, *Magogo & Glover* 491!; Kilifi District: Kaya Jibana, Aug. 1996, *Luke et al.* 4528! & Kaya Kauma, 14 May 1998, *Luke* 5276

Tanzania. Lushoto District: Sigi–Kisiwani, 4 June 1941, *Omari bin Bakari* AH9787!; Morogoro District: Kimbosa Forest Reserve, *Mwasumbi & Harris* 2420!; Pemba: Ngezi Forest, Dec. 1989, *Beentje* 4319!

Distr. **K** 7; **T** 3, 6; **P**; not known elsewhere

Hab. Moist forest; 0–800 m

Uses. None recorded

10. **S. hispidus** *DC.* in Bull. Soc. Philom. Paris 64: 122 (1802); Stapf in F.T.A. 4, 1: 174 (1902); Gilg in Monogr. Afr. Pfl.-Fam. & Gatt. 7: 35, t. 2 (1903); T.T.C.L.: 56 (1949); Huber in F.W.T.A. ed. 2, 2: 70 (1963); Beentje in Meded. Landbouwhogeschool 82 (4): 85, t. 21 (1982). Type: Sierra Leone, sine loc., *Smeathmann* s.n. (G-DC, holo., BM!, K!, P!, UPS, iso.)

Sarmentose shrub 1.5–5 m high or liana to 100 m long; deciduous; trunk with dark grey bark; branches dark brown or blackish, lenticellate; branchlets reddish brown, densely hispid; latex clear, reddish or white. Leaves opposite or rarely ternate, petiolate, blade glossy green above, dull and paler beneath, ovate, elliptic or obovate, 3–15 cm long, 1.5–8 cm wide, base rounded or subcordate, apex acuminate, sparsely to densely hispid; petiole 1–5 mm long. Inflorescence terminal or in forks, 1–72-flowered, sessile or pedunclate, lax or congested, hispid in all parts; pedicel 5–32 mm long. Flowers with sepals unequal, the outer ovate, the inner linear, 13–35 mm long, acute; corolla tube white, turning yellow and then orange, red- to purple-spotted inside, 11–22 mm long, corolla lobes creamy turning orange, ovate, 3–10 mm long, 3–8 mm wide, narrowing into 15–23 cm long pendulous tails, these yellow, turning reddish distally, corona lobes yellow to purple-spotted, 1–3 mm long. Fruit dark brown, hard, with mericarps opposite-divergent or at an obtuse angle, almost cylindrical, 24–48 cm long, long-tapering and abruptly widening into a terminal knob, lenticellate, sulcate and hispid to glabrescent; seed 10–18 mm long, with a stalked coma 5–12 cm long.

Uganda. Ankole District: Kashoya–Kitomi forest, Aug. 1936, *Eggeling* 3210!; Mengo District: Kitubulu, Nov. 1938, *Chandler* 2493!; Masaka District: Misozi, *Bagshawe* 128!

Tanzania. Mwanza District: Rubondo I., near Kageye, 18 Oct. 1985, *Fitzgibbon & Barcock* 102!; Kigoma District: Kasikati basin, *Itani* 49!

Distr. **U** 2, 4; **T** 1, 4; West Africa from Senegal and Guinea to Central African Republic and northern Angola

Hab. Moist forest; 1100–1600 m

Uses. Source of arrow poison; source of heart medicine (strophanthin)

SYN. *S. hispidus* DC. var. *bosere* De Wild., Miss. Laurent: 546 (1907). Type: Congo (Kinshasa), Eala, *Malchair in herb. Laurent* 1273 (BR!, holo.)

11. **S. kombe** *Oliv.* in Hook., Ic. Pl. 3 (1): 79, t. 1098 (1871); Stapf in F.T.A. 4, 1: 173 (1902); Gilg in Monogr. Afr. Pfl.-Fam. & Gatt. 7: 36, t. 3 (1903); T.T.C.L.: 56 (1949); Codd in F.S.A. 26: 290 (1963); Verdcourt & Trump, Comm. Poison. Pl. E. Afr.: 138, t. 10/e–g (1969); Beentje in Meded. Landbouwhogeschool 82 (4): 96, t. 24–25 (1982) & in F.Z. 7 (2): 474, t. 112 (1985); K.T.S.L.: 485, map (1994). Type: Malawi, Manganja Hills, *Meller* s.n. (K!, holo.)

Sarmentose shrub 1–3.5 m high or liana to 20 m long, deciduous, flowers appearing with the leaves, trunk to 10 cm in diameter, bark reddish brown or grey-brown; branches scabrid; branchlets densely hispid; latex clear, white or yellow. Leaves opposite; blade often convex, ovate or elliptic, 8–24 cm long, 5–17 cm wide, base cuneate, rounded or subcordate, apex obtuse, acute or acuminate, densely hispid on both surfaces but glabrescent above; petiole 1–5 mm long. Inflorescence 1–12-flowered, pedunculate, fairly congested, densely hispid in all parts; pedicels 3–20 mm long. Flowers fragrant; sepals narrowly ovate or linear, 9–27 mm long, acute; corolla white turning yellow, red-spotted inside; corolla tube 13–24 mm long, corolla lobes ovate, 3–16 mm long, 4–9 mm wide, narrowing into the 9–20 cm long pendulous tails, corona lobes 1–3 mm long. Fruit hard, the mericarps opposite-divergent, cylindrical or nearly so, 15–47 cm long, 1–2.6 cm in diameter, tapering to a knob, lenticellate, hispid and glabrescent; seeds 11–21 mm long, densely pubescent, with a stalked coma 6–14 cm long.

KENYA. Lamu District: Mundane Range, 5 Apr. 1980, *Gilbert & Kuchar* 5929!; Kilifi District: Kaya Kivara, 7 July 1987, *Robertson & Luke* 4730!; Kwale District: Muhaka Forest, 14 Apr. 1977, *Gillett* 21065!
TANZANIA. Tanga District: Mtotohovu, 10 Sept. 1951, *Greenway* 8704!; Uzaramo District: Pande Forest Reserve, *Harris et al.* 3623!; Kilwa District: Malemba thicket, 14 Dec. 1976, *Vollesen* MRC4221!
DISTR. **K** 7; **T** 3, 6, 8; Zambia, Malawi, Zimbabwe, Mozambique, Namibia, Botswana, NW South Africa
HAB. Coastal evergreen forest and thicket, dense bushland; 0–850 m
USES. Seeds and roots source of arrow poison; source of heart medicine (strophanthin)

SYN. *S. hispidus* DC. var. *kombe* (Oliv.) Holmes in Pharm. J. Trans. ser. 3, 21: 223 (1890)

12. **S. bequaertii** *Staner & Michotte* in B.J.B.B. 13: 53 (1934); Beentje in Meded. Landbouwhogeschool 82 (4): 45, t. 9 (1982). Type: Congo (Kinshasa), Kivu, Masisi, *Bequaert* 6376 (BR!, holo. & iso.)

Liana to 10 m long; latex white. Leaves opposite, blade elliptic or slightly obovate, 4–11.5 cm long, 1.5–4.5 cm wide, base cuneate, apex acuminate, glabrous; petiole 4–12 mm long. Inflorescence 1–6-flowered; peduncle 2–12 mm long; pedicels 4–11 mm long. Sepals equal, erect or spreading, ovate, 5–12 mm long, 1.5–3 mm wide, acute, glabrous; corolla white turning yellow with violet tinge outside and violet streaks inside; tube 15–25 mm long, corolla lobes ovate, 5–7 mm long, 3.5–7 mm wide, narrowing into 2–6 cm long spreading or pendulous tails, corona lobes 4–6.5 mm long. Fruit woody, hard, the mericarps cylindrical or nearly so, 18–21 cm long, 1.4 cm in diameter, tapering to a knob or less often into an obtuse tip, lenticellate, glabrous; seeds 15–17 mm long, densely puberulous, with a stalked coma 4–4.5 cm long.

UGANDA. Kigezi District: Bwindi Forest, Aug. 1998, *Eilu* 318! & 322!
DISTR. **U** 2; E Congo (Kinshasa), Rwanda
HAB. Moist forest; ? 1400 m
USES. None recorded

25. FUNTUMIA

Stapf in Proc. Linn. Soc. 7: 2, 3 (1899); Zwetsloot in Meded. Landbouwhogeschool
81 (16): 10–32 (1981)

Evergreen trees or shrubs, with white latex; trunk mostly straight, cylindrical. Leaves opposite, those of a pair connate into a short ochrea, with many small colleters in two or three rows in the axils; blade with domatia beneath; secondary veins prominent beneath. Inflorescence terminal and axillary, cymose, congested; bracts ovate or elliptic with colleters within; peduncle short. Flowers 5-merous, actinomorphic, fleshy, fragrant. Sepals free, thick, with colleters within. Corolla tube ventricose at the middle, throat thickened; lobes overlapping to the right; stamens included; filaments short or absent; anthers 2-celled, introrse; connective coherent with the pistil head; ovary of two almost free carpels, united at the base of the style; pistil head ovoid, stigma conical; disk 5-lobed; ovules pendulous, 200–350 in each carpel. Fruit of two separate woody follicles, connate at the base, flattened adaxially and dehiscent; seeds slender, fusiform, with a hairy coma at the apex.

Two species in Africa.

1. Domatia usually absent; corolla pale green to creamy,
     puberulous or glabrous; lobes obliquely ovate or narrowly
     oblong sometimes auriculate; ovary pubescent · · · · · · · · ·   1. *F. africana*
   Domatia often present; corolla white, glabrous; lobes
     triangular to ovate and distinctly auriculate; ovary glabrous ·   2. *F. elastica*

1. **F. africana** (*Benth.*) *Stapf* in Proc. Linn. Soc. 7: 2, 3 (1899) & in Hook. Ic. Pl. 7, 4: 2696, 2697 (1901); Huber in F.W.T.A. ed. 2, 2: 74 (1963); Hamilton, Field Guide Uganda For. Trees: 164 (1981); Zwetsloot in Meded. Landbouwhogeschool 81 (16): 16, t. 3, photo 4, 6, 7, 8 (1981) & in F.Z. 7, 2: 485, t. 115 (1985); K.T.S.L.: 479, ill. (1994). Type: Equatorial Guinea, Bioco [Fernando Po], Bagru R., *Mann* 817 (K, lecto., GH, P, isolecto., chosen by Zwetsloot)

Tree 8–30 m high; latex not easily coagulating; trunk up to 50 cm in diameter; branchlets glabrous or minutely pubescent; bark smooth, sometimes with a few orbicular lenticles, greenish brown to grey, mottled; wood light, soft. Leaves petiolate; blade ovate or elliptic, 5–32 cm long, 1.7–17 cm wide, base cuneate, apex shortly acuminate, often puberulous on midrib; domatia if present of a tuft of straight hairs; petiole 3–15 mm long, glabrous or minutely pubescent. Inflorescence 2–2.5 cm long, 3–40-flowered; bracts 0.9–2 mm long, obtuse or acute; pedicels 3–15 mm long, glabrous or puberulous. Flowers with sepals triangular to broadly ovate, 1.5–4 mm long, with 5–50 colleters in a single flower; corolla tube very pale green to creamy, 5.8–10 mm long, glabrous to puberulous outside; lobes creamy, obliquely ovate to narrowly oblong, sometimes auriculate, 5–17 mm long, 2–5 mm wide, obtuse or acute at the apex; disk 5-lobed; pistil 4–7 mm long; ovary pubescent at the apex; style 1.2–3 mm long. Fruits grey-brown, fusiform, 8.4–32 cm long, 1.4–2 cm in diameter; seed 3.5–7.5 cm long, 0.2–0.6 cm wide, coma 1.6–6 cm long; glabrous basal portion of beak 0.1–1.1 cm, hairs 3.2–9 cm long. Fig. 29 (p. 87).

UGANDA. Toro District: Itwara forest, Feb. 1943, *Thompson* E5230!; Mbale District: Balakeki, near Bubulu, Sept. 1931, *Harris* 133!; Masaka District: NW side of Lake Nabugabo, 9 Oct. 1953, *Drummond & Hemsley* 4728!
KENYA. Nandi District: Kaimosi Forest, 22 Dec. 1971, *Kokwaro* 10!; North Kavirondo District: Kakamega, near Yala R., Apr. 1934, *Dale* 802!; Kwale District: Mwele Mdogo Forest, Shimba Hills, 5 Jan. 1988, *Luke* 890!

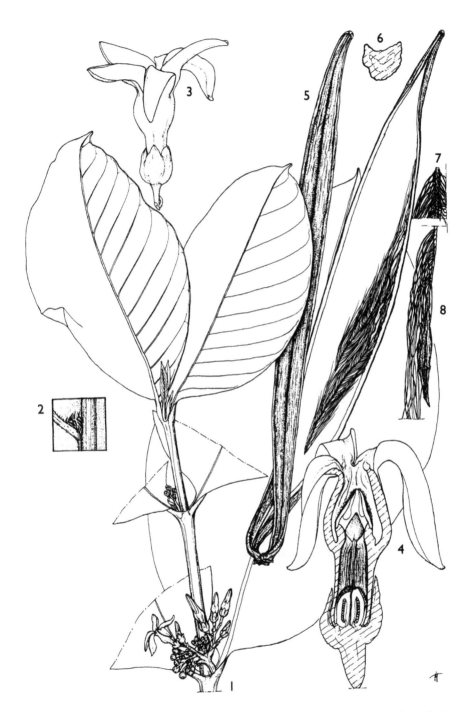

FIG. 29. *FUNTUMIA AFRICANA* — **1**, flowering branch × ²/₃; **2**, domatium in the axil of a secondary vein, × 4; **3**, flower, × 2; **4**, LS flower, × 4; **5**, opened fruit, × ²/₃; **6**, TS fruit, × ²/₃; **7**, detail of haired beak, × 2; **8**, detail of seed, × 2. 1–4 from *Zwetsloot* 28; 5–8, from *Zwetsloot* 21. From Meded. Landbouwhogeschool 81(16), drawn by Y.F. Tan, reproduced by permission.

Tanzania. Mwanza District: Rubya Forest Reserve, Apr. 1949, *Watkins* 240!; Lushoto District: Longuza Forest Reserve, 7 Mar. 1962; *Mgaza* 462!; Kigoma District: Kaskati basin area, 1966, *Itani & Izawa* 50!; Pemba: Ngezi Forest, July 1901, *Lyne* 118

Distr. U 2–4; **K** 3, 5, 7; **T** 1, 3, 4, 6–8; **Z**, **P**; tropical Africa from Senegal to Congo (Kinshasa) to Angola and S to Zimbabwe and Mozambique

Hab. Moist forest, riverine or swamp forest; 30–1600 m

Uses. Latex used as bird lime; wood for construction, furniture and small implements; minor medicinal use for burns, constipation, incontinence

Syn. *Kickxia africana* Benth. in G.P. 2: 1276 (1879)
   K. latifolia Stapf in K.B. 1898: 307 (1898); I.T.U. ed. 2: 28 (1952). Type: Congo (Kinshasa), Prov. Equateur, Bangala, *Dewèvre* 867 (BR, lecto., K, iso.)
   Funtumia latifolia (Stapf) Stapf in Proc. Linn. Soc., 1899: 2, 3 (1899); T.T.C.L.: 50 (1949); I.T.U. ed. 2: 28 (1952); K.T.S.: 46 (1961)
   Kickxia congolana De Wild. in Rev. Cult. Colon. 7: 745 (1900). Type: Congo (Kinshasa), Bas-Zaire, Kisantu, *Gillet* 387 (BR, holo., K, iso.)
   K. gilletii De Wild. in Rev. Cult. Colon. 7: 744 (1900). Type: Congo (Kinshasa), Bas-Zaire, Kisantu, *Gillet* 886 (BR, holo., K, iso.)
   K. scheffleri K. Schum. in N.B.G.B. 3 (21): 81 (1900). Type: Tanzania, Lushoto District, Derema, *Scheffler* 176 (B, holo., BM, E, EA, K, P, Z, iso.)
   Funtumia congolana (De Wild.) Jumelle, Plantes Caoutchouc: 381 (1903)
   F. gilletii (De Wild.) Jumelle, Plantes Caoutchouc: 381 (1903)

2. **F. elastica** (*Preuss*) *Stapf* in Proc. Linn. Soc. 7: 2, 3 (1899); T.T.C.L.: 50 (1949); I.T.U. ed. 2: 27 (1952); Huber in F.W.T.A. ed. 2, 2: 74, t. 216 (1963); Zwetsloot in Meded. Landbouwhogeschool 81 (16): 25, t. 4, photo 3, 5, 9 (1981). Type: Cameroon, Yaounde, *Preuss* 1381 (B†, holo., K, lecto., BM, BR, E, MO, P, W, iso., chosen by Zwetsloot)

Tree up to 35 m high; white latex coagulating easily; trunk up to 50 cm in diameter; branchlets glabrous. Leaves petiolate; blade 6–27 cm long, 1.5–10 cm wide, base rounded or cuneate, apex acuminate, glabrous except for a few hairs beneath on midrib; domatia often present, consisting of pits with a ciliate margin; petiole 2–15 mm long, glabrous. Inflorescence 2.2 cm long, 3–35-flowered; pedicel 2–8 mm long, glabrous. Flowers with sepals elliptic to broadly ovate, 3–5 mm long; corolla tube very pale green to white, 5.5–10.5 mm long, 3–5.5 mm wide, glabrous outside; lobes white, triangular to ovate, distinctly auriculate, 3–7 mm long, 2–4 mm wide, glabrous on both sides; disk with truncate lobes; pistil 4–6 mm long; ovary glabrous; style 1.5–3 mm long, glabrous. Fruit almost clavate, 8–19 cm long, 2–4 cm in diameter, subacute or acute at the apex; seeds 4.5–7 cm long, 0.3–0.4 cm wide; coma 3–5 cm long, glabrous basal portion 0.5–1.3 mm long, hairs 4.5–7 cm long.

Uganda. Toro District: Semliki Forest, Bwamba, Feb. 1943, *Thompson* s.n.!; Bunyoro District: Bujenje, Budongo Forest, 19 Aug. 1971, *Synnott* 642! & Muhimbi sample plot on Sonso R., Nov. 1935, *Eggeling* 2319!

Tanzania. Tanga District: Muheza, along road to Amani, *Perdue & Kibuwa* 8484 & Amani, 5 Dec. 1910, *Rohrer* s.n.

Distr. U 2, 4; **T** 3; tropical Africa from Senegal to Congo (Kinshasa) and Sudan

Hab. Moist forest; 1050–1200 m

Uses. Valuable rubber tree; wood for implements; latex and bark as minor medicinal use against impotence, piles and jaundice

Syn. *Kickxia elastica* Preuss in N.B.G.B. 2 (19): 353–360, t. 1 (1899)

Note. Cultivated in Amani.

26. MASCARENHASIA

A. DC., Prodr. 8: 487 (1844); Leeuwenberg in Wageningen Agric. Univ. Pap. 97, 2: 21–52 (1997)

Shrubs or trees, mostly with white latex in all parts; branchlets terete. Leaves opposite, sometimes deciduous; ochreae not or slightly widened into intrapetiolar stipules. Inflorescence axillary and/or terminal; peduncle subsessile; bracts sepal-like, persistent. Flowers fragrant or not. Sepals free. Corolla tube cylindrical in the lower part, abruptly widening just below the insertion of the stamens into a campanulate or narrowly urceolate apical part; lobes free to halfway connate, overlapping to the right; stamens included, barely exserted; anthers with pubescent connective, coherent with a circular patch with pistil head; ovary of 2 separate carpels; disk present, 5-parted. Fruit of 2 linear cylindrical follicles, abaxially and longitudinally dehiscent; seeds dark brown, with a coma longer than the grain, directed towards the apex of the follicle.

Eight species, seven endemic to Madagascar, the eighth also in E & SE Africa.

M. arborescens A. DC., Prodr. 8: 488 (1844); Huber in F.W.T.A. ed. 2, 2: 72 (1963); Kupicha in F.Z. 7 (2): 487, t. 116 (1985); K.T.S.L.: 482, ill. (1994); Leeuwenberg in Wageningen Agric. Univ. Pap. 97, 2: 24, t. 3, phot. 1 (1997). Type: Madagascar, Bombetoka bay, *Bojer* s.n. (G-DC, holo.)

Tree or shrub 1.5–15(–20) m high, with white latex in all parts; trunk 3–40 cm in diameter; bark smooth or rough, often peeling off in large scales; branchlets glabrous. Leaves petiolate; blade elliptic or obovate, 4–18 cm long, 2–6.5 cm wide, acuminate to retuse at the apex, cuneate at the base, revolute, glabrous, petiole 1–8 mm long, glabrous. Inflorescence fasciculate, 2–4 cm long, 2–6 cm wide, 1–15-flowered; peduncle 2–5 mm long, glabrous; pedicels 7–12 mm long, glabrous to puberulous. Flowers with sepals ovate or triangular, 1.5–5.5 mm long, 1.5–4 mm wide, with a row of 7–10 colleters within; corolla white with yellow streak inside, pubescent inside on lobes, up to insertion of stamens; tube 5–14 mm long, widest above the base and around the stamens; lobes ovate, 4–15 mm long, 2–14 mm wide, apiculate, acuminate or acute, densely hairy; stamens with apex 0.2–1.8 mm below the mouth of the corolla tube, inserted 5–8 mm from the base; disk (1) 2–3 mm, usually higher than the ovary; ovary cylindrical, pubescent at the apex, rarely entirely glabrous; ovules 30–60 in each carpel; style pubescent; pistil head almost conical. Follicles dark brown, 6–17.5 cm long, 0.6–2 cm wide; seed 10–14 mm long, dark brown; coma 10–20 mm long. Fig. 30 (p. 90).

KENYA. Kwale District: Buda Forest, Nov. 1936, *Dale* 1053! & Marere water works, 14 Jan. 1970, *Perdue & Kibuwa* 10220!; Kilifi District, Pangani, crossing of Lwandani stream, 17 Feb. 1977, *Faden et al.*, 77/532!
TANZANIA. Pangani District: Pangani, 1973, *Bond* 78!; Morogoro District: Turiani Falls, 10 Mar. 1953, *Drummond & Hemsley* 1473!; Uzaramo District: Pugu Hills, 14 Nov. 1971, *Hansen* 521!; Pemba: Mlimi, Oct. 1951, *R.O. Williams* 99
DISTR. **K** 7; **T** 3, 6; **Z**, **P**; Congo (Kinshasa), Malawi, Mozambique, Zimbabwe, Comoro Islands and Madagascar
HAB. On river banks in forest or gallery forest; 0–750 m
USES. Yields a good rubber

SYN. *M. variegata* Britt. & Rendle in Trans. Linn. Soc. 2, 4: 26 (1894). Type: Malawi, Mt Mlanje, *Whyte* 108 (BM, holo., G, K, Z, iso.)
    M. fischeri K. Schum. in P.O.A. C: 318 (1895). Type: Tanzania, *Fischer* 322 (B†, holo.)
    M. elastica K. Schum. in N.B.G.B. 2: 270 (1899); K.T.S.: 47 (1961). Type: Tanzania, Uzaramo District, between Dar es Salaam and Mpafu [Mbaffu], near Vikondo, *Stuhlmann* s.n. (B†, holo.)
    Lanugia variegata (Britt. & Rendle) N. E. Br. in Torreya 27: 53 (1927); T.T.C.L.: 53 (1949)

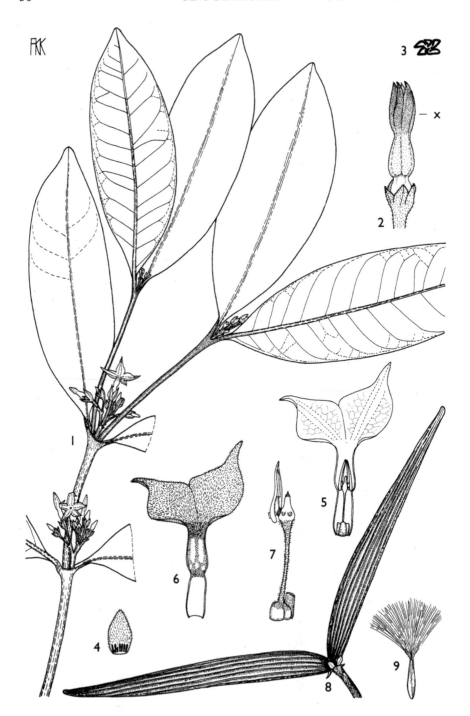

Fig. 30. *MASCARENHASIA ARBORESCENS* — **1**, habit × ²/₃; **2**, flower bud, × 2; **3**, CS bud at level X, × 2; **4**, calyx lobe, adaxial view, × 4; **5**, part of flower, × 2; **6**, corolla, × 2; **7**, one stamen gynoecium and disk scales, × 4; **8**, fruit, × ²/₃; **9**, seed, × ²/₃. 1 from *Torre* 654; 2–3 from *Rail* 1/55; 4–7 from *Mendonca* 2478; 8 from *Barbosa* 2115; 9 from *Andrada* 1911. From F.Z., drawn by F.K. Kupicha.

Fig. 31. *HOLARRHENA PUBESCENS* — **1**, flowering branch × ²/₃; **2**, part of branch, × ²/₃; **3**, bark, × ²/₃; **4**, flower, × 2; **5**, part of flower, × 2; **6**, sepal inside with colleters, × 6; **7**, sepal inside with colleter, × 8; **8**, flower with petals removed, × 4; **9**, part of corolla with stamens, × 8; **10**, fruit, × ²/₃; **11**, seed × ²/₃. 1 from *Pawek* 7734; 2 from *Thulin & Mhoro* 1256; 3–4 from *Leeuwenberg* 10867; 5-6, 8-9, from *Leeuwenberg* 10866; 7 from *Squires* 778; 10 from *Lawton* 1212; 11 from *Harmand* 23. From Meded. Landbouwhogeschool 81(2), drawn by J. Williamson, reproduced by permission.

27. HOLARRHENA

R. Br. in Mem. Wern. Soc. 1: 62 (1811); de Kruif in Meded. Landbouwhogeschool
81 (2): 4–36 (1981)

Deciduous shrubs or trees up to 25 m high, with abundant white latex in all parts.
Leaves (sub-)opposite, petiole glandular near the base. Inflorescences terminal,
cymose, many-flowered; bracts without colleters. Flowers fragrant. Sepals persistent,
valvate, with 0–10 colleters at the base. Corolla salver-shaped; tube cylindrical,
pubescent at the throat; lobes white, overlapping to the right; stamens inserted at the
base of the corolla tube; filaments short, pubescent at the base; pistil glabrous or
pubescent; disk absent; ovary ovoid, of two carpels, connate at the extreme base,
placentas adaxial; ovules numerous, in 4–8 rows; pistil head ovoid; stigmoid apex
bifid. Fruit of two thin-walled slender follicles, dehiscent; seeds glabrous, with a
dense tuft of hairs at the apex, 3–6 × as long as the seeds; embryo large, with a scanty
endosperm; cotyledons large, complicate.

Four species in Asia and Africa.

H. pubescens (*Buch.-Ham.*) *G. Don*, Gen. Syst.: 78 (1837); Codd in F.S.A. 26: 263
(1963); de Kruif in Meded. Landbouwhogeschool 81 (2): 17, t. 5 (1981) & in F.Z. 7,
2: 456 (1985); K.T.S.L.: 480, ill., map (1994). Lectotype: Burma, Ava, Conoungaklay,
Buchanan-Hamilton s.n. (BM, lecto., chosen by de Kruif)

Shrub or tree 0.6–18 m high, deciduous; trunk 12–25 cm in diameter; bark
smooth, lenticellate or rough and corky, pale to dark grey; wood soft; branchlets
pubescent, rarely glabrous. Leaves petiolate; blade ovate to elliptic, 1.7–23 cm long,
1.3–11.5 cm wide, acuminate or acute at the apex, cuneate or rounded at the base,
pubescent to glabrous; petiole 2–12 mm long, often pubescent. Inflorescences
seemingly axillary or leaf-opposed, sometimes terminal, mostly pubescent; bracts
1.5–4.5 mm long; peduncle 0.9–1.7 cm long; pedicels 0.3–3 cm long. Flowers with
sepals elliptic to linear, 2–12 mm long; corolla tube white, 9–19 mm long, outside
with ordinary and glandular hairs, rarely glabrous; lobes white, narrowly triangular,
10–24(–30) mm long, 3–8 mm wide, pubescent and ciliate or glabrous; stamens
inserted at 1.5–2.5 mm from the base of the tube; pistil 1.8–3.1 mm long; style
0.3–1.2 mm long. Fruits pale grey to dark brown, dotted with white, 20–38 cm long,
2–9 mm in diameter; seeds 9–16 mm long; tufts of hairs 25–45 mm long. Fig. 31 (p. 91).

KENYA. Teita District: Wassesi, Tsavo National Park East, 5 Dec. 1966, *Greenway & Kanuri*
12653!; Kilifi District: Arabuko Sokoke Forest, Jan. 1937, *Dale* 1073!; Lamu District: Witu
Forest, 4 Aug. 1988, *Gachathi* 50/88!
TANZANIA. Mwanza District: Mwanza, Capri point, Jul. 1951, *Eggeling* 6262!; Mpanda District:
Katavi Game Reserve, 10 Nov. 1963, *Carmichael* 1007!; Kilosa District: Ruaha Gorge, 35 km E
of Mbuyuni, 3 Feb. 1974, *Barley & Carter* 16419!
DISTR. **K** 7; **T** 1, 3–8; Congo (Kinshasa), Zambia, Malawi, Mozambique, Zimbabwe, South
Africa; S Asia
HAB. Deciduous forest, riverine forest, woodland, evergreen bushland; 0–1250 m
USES. Roots used against influenza and venereal diseases; leaf infusion against stomach pains

SYN. *Echites pubescens* Buch.-Ham. in Trans. Linn. Soc. London 13: 521–525 (1822), *non* Hooker
& Arnott (1841)
　　Holarrhena febrifuga Klotzsch in Reise Mossamb.: 277 (1862); P.O.A. C: 137 (1895); Stapf in
　　F.T.A. 4, 1: 107 (1902); T.T.C.L.: 50 (1949); K.T.S.: 47 (1961). Type: Mozambique, Tete,
　　Sena, *Peters* s.n. (B†, holo.); neotype: Mozambique, Tete, Nyampanda, *Gillett* 17510
　　(WAG, neo.; K, SRGH, isoneo., chosen by de Kruif)
　　H. glabra Klotzsch in Reise Mossamb.: 279 (1862). Type: Mozambique, Tete, *Peters* s.n. (B†,
　　holo.); neotype: Mozambique, Monapo, km 11 on Namialo–Meserepane road, *Torre &
　　Paiva* 9272 (LISC, neo., chosen by de Kruif)

H. tettensis Klotzsch in Reise Mossamb.: 278 (1862). Type: Mozambique, Tete, Sena, *Peters* s.n. (B†, holo.); neotype: Mozambique, Namagoa, Milanje road, 30–40 km from Mocuba, *Faulkner* 488 (BR, neo., COI, K, S, SRGH, isoneo., chosen by de Kruif)

H. febrifuga Klotzsch var. *glabra* Oliv. in Trans. Linn. Soc. 29 (3): 108 (1875). Type: Tanzania, Morogoro/Rufiji District, Ukufu [Khutu], Mbwiga, *Speke & Grant* s.n., Nov. 1860 (K, holo.)

H. fischeri K. Schum. in P.O.A. C: 316 (1895). Type: Tanzania, Singida District, Unyamwezi, Ussure [Usure], *Fischer* 378 (B†, holo., K, fragment)

H. glaberrima Markgr. in Mitt. Bot. Staatss., München 1 (1): 28 (1950). Type: Tanzania, Uluguru District, Morogoro, Uluguru Mts, *Schlieben* 3326 (M, holo., B, BM, G, LISC, P, PRE, Z, iso.)

28. ALAFIA

Thouars, Gen. Nov. Madag.: 11 (1806); Leeuwenberg in K.B. 52 (4): 769–839 (1997)

Lianas or climbing shrubs; white or clear latex in all parts (except in *A. zambesiaca* and occasionally *A. schumannii*); branches mostly lenticellate; branchlets terete, glabrous or sometimes puberulous. Leaves opposite, petiolate; ochreae often widened into intrapetiolar stipules. Inflorescence terminal, sometimes also axillary, cymose, few- or many-flowered, lax to very dense; peduncle mostly short; flowers open during the day, mostly fragrant. Sepals connate at the extreme base, ciliate, mostly with colleters within. Corolla with variously shaped heads in mature bud more or less characteristic for the species, mostly pubescent below the insertion of the stamens and interrupted between the filament ridges; lobes overlapping to the right in bud, ciliate; tube shorter or longer than the lobes; lobes entire to sinuate-crenate, spreading or recurved; stamens with their tips reaching mouth of corolla tube; anthers sessile, sterile at the apex and below, pubescent below, coherent with the pistil head by a clavuncle; ovary of two separate carpels (united in *A. multiflora*), pubescent at the apex; disk absent; pistil of a basal cone or cylinder, usually with a ring at the apex and a bilobed stigmoid apex. Fruit of two separate, thin walled, long and linear follicles (1 syncarpous in *A. multiflora*), dehiscent on adaxial side; seeds with a short wing at the base and a hairy coma longer than the grain attached at the apex or on a beak at the apex.

23 species, 15 in Africa and 8 in Madagascar.

1. Leaves > 15 cm long, densely pubescent beneath; branchlets puberulous · 2. *A. erythrophthalma*
   Leaves ≤ 15 cm long, glabrous; branchlets glabrous (or occasionally puberulous in *A. microstylis*, which also has puberulous leaves) · 2
2. Leaf with 11–14 pairs of main lateral veins; corolla lobes spoon-shaped · 4. *A. microstylis*
   Leaf with 4–9 pairs of main lateral veins; corolla lobes obliquely ovate to obovate or dolabriform · · · · · · · · · · · · · · · · 3
3. Inflorescence lax, few-flowered (up to 17) · 4
   Inflorescence dense, many-flowered · 5
4. Corolla lobes 4.5–12 mm long, less than $1\frac{1}{2}$ × as long as wide · 1. *A. caudata*
   Corolla lobes 9–9.3 mm long, more than $2\frac{1}{2}$ × as long as wide · 6. *A. zambesiaca*
5. Leaves obovate, rounded to shortly acuminate at the apex; corolla lobes dolabriform · · · · · · · · · · · · · 3. *A. lucida*
   Leaves elliptic; corolla lobes obliquely (ob-)ovate · 6

6. Leaves bluntly acuminate to obtuse at the apex; corolla
   orange to yellow or reddish outside; fruit follicles
   20–60 cm long · 5. *A. orientalis*
   Leaves apiculate with acute acumen; corolla white; fruit
   follicles 40–110 cm long · · · · · · · · · · · · · · · · · · 7. *A. schumannii*

1. **A. caudata** *Stapf* in K.B. 1894: 123 (1894); Pichon in B.J.B.B. 24: 164, t 3/c–d (1954), pro parte; Kupicha in B.J.B.B. 51: 160 (1981) & in F.Z. 7, 2: 490 (1985); K.T.S.L.: 475, ill. (1994); Leeuwenberg in K.B. 52 (4): 781, t.3 (1997). Lectotype: Angola, Golungo Alto, Alto Queta Mts, *Welwitsch* 5955 (K, lecto., BM, BR, C, P, iso., chosen by Kupicha)

Liana 3–25 m high with white latex; branches dark brown with pale brown lenticels; branchlets glabrous. Leaves petiolate; blade (narrowly) elliptic, 3–9 cm long, 1–4 cm wide, acuminate or caudate at the apex, cuneate or rounded at the base, glabrous, with 4–9 pairs of secondary veins forming an angle of 45–60° with the midrib; petiole 1–5 mm long, glabrous. Inflorescence terminal and often also axillary, few-flowered, glabrous; peduncle 3–16 mm long; pedicels 3–10 mm long. Flowers fragrant; sepals ovate, 1.2–2.2 mm long, 0.8–1.4 mm wide, obtuse, glabrous, with one colleter near the edge; corolla white, tube often partly pink outside; mature bud with an ovoid head about as long as the tube; tube 5–9 mm long, widest at the insertion of the stamens; lobes obliquely oblong or obliquely ovate, 4.5–12 mm long, 2.5–8 mm wide, rounded, spreading or recurved; stamens with apex 1.2 mm below to 1 mm above the mouth of the corolla tube, inserted at 2–5 mm from the base; anthers 2.8–4.5 mm long; style 2.2–4 mm long; pistil head 0.5–1 mm long, stigmoid apex minute; ovules ± 100 in each carpel. Fruit brown, 15–40 cm long, 0.5–1 cm wide, glabrous; seed 15–25 mm long; coma 2.5–3.5 cm long.

KENYA. Kwale District: Dzombo Mt, 8 Apr. 1968, *Magogo & Glover* 791! & Buda Mafisini Forest Reserve, 25 Feb. 1989, *Luke & Robertson* 1680!; Kilifi District: Mangea Hill, 16 Feb. 1988, *Luke & Robertson* 985!
TANZANIA. Lindi District: Rondo Forest Reserve, 9 Feb. 1991, *Bidgood et al.*, 1423! & Jun. 1944, *Gillman* 1555!; Newala District: Kitangari, May 1943, *Gillman* 2564!; Pemba: Ngezi Forest, Dec. 1983, *Rodgers et al.* 2722
DISTR. **K** 7; **T** 8; **P**; Cameroon, Gabon, Congo (Kinshasa), Angola, Mozambique and Zimbabwe
HAB. Forest, gallery forest or bushland; 0–800 m
USES. None recorded

SYN. *A. butayei* Stapf in F.T.A. 4, 1: 199 (1902). Type: Congo (Kinshasa), Bandundu, between Dembo & Kwango R., *Butaye* 1491 (BR, holo., K, iso.)
　　A. caudata Stapf subsp. *latiloba* Kupicha in B.J.B.B. 51: 161, t. 4 (1981). Type: Tanzania, Newala, *Hay* 77 (K, holo., B, BR, S, iso.)

2. **A. erythrophthalma** (*K. Schum.*) *Leeuwenberg* in Novon 6: 271 (1996) & in K.B. 52 (4): 785, t. 4 (1997). Type: Cameroon, Yaoundé, *Zenker & Staudt* 701 (B†, holo., K, lecto., BM, iso., chosen by Leeuwenberg)

Large liana up to 15 m high, with white latex; trunk 3–10 cm in diameter; bark smooth; branches dark brown with cream lenticels; branchlets puberulous. Leaves petiolate; blade elliptic, rarely obovate, 15–32 cm long, 4–15 cm wide, apiculate at the apex, rounded or less often cuneate at the base, glabrous above, densely pubescent beneath, with 5–20 pairs of rather straight secondary veins forming an angle of 45–80° with the midrib; petiole 3–10 mm long, puberulous. Inflorescence terminal, 4–12 cm long, many-flowered, dense, puberulous; peduncle 5–30 mm long; pedicels 1–3 mm long. Flowers fragrant; sepals ovate, 1.5–3 mm long, 1.3–2 mm wide, obtuse, with one large colleter near the edge; corolla yellow, orange, orange-brown or cream, red in the throat and often also outside the tube; mature bud with an ellipsoid or (narrowly) ovoid head about 0.5–1 × as long as the tube, ciliate; tube 7–11.5 mm long, 1.5–2 mm wide; lobes obliquely obovate, 5–9 mm long, 4–6.5 mm wide, rounded; stamens inserted 3.5–5.5 mm from the base; style 2–3.1 mm long; ovules 50–100 in each carpel. Fruit dark brown, 26–62 cm long, 0.7–1 cm wide, pubescent; seed up to 25 mm long; coma 3–7 cm long.

UGANDA. Bunyoro District: Budongo Forest, 8 Mar. 1973, *Synnott* 1449! & Bugoma Forest, 1905, *Dawe* 730; ?Mengo District: Buvuma Is., c. 1905, *Christy* s.n.
DISTR. U 2, 4; Nigeria, Cameroon, Equatorial Guinea, Gabon, Congo (Brazzaville), Central African Republic and Congo (Kinshasa), Angola (Cabinda)
HAB. Moist forest; 1050–1100 m
USES. None recorded

SYN. *Tabernaemontana erythrophthalma* K. Schum. in E.J. 23: 224 (1896)
Alafia grandis Stapf in F.T.A. 4, 1: 196 (1902); Pichon in B.J.B.B. 24: 198, t. 5/c–d (1954); Huber in F.W.T.A. ed. 2, 2: 73 (1963). Lectotype: Cameroon, Yaoundé, *Zenker & Staudt* 213 (K, lecto., chosen by Pichon)

3. **A. lucida** *Stapf* in K.B. 1894: 122 (1894); T.T.C.L.: 48 (1949); Pichon in B.J.B.B. 24: 181, t. 4/c–d (1954); Huber in F.W.T.A. ed. 2, 2: 73 (1963); Leeuwenberg in K.B. 52 (4): 794, t. 7 (1997). Type: Equatorial Guinea, Rio Muni, *Mann* 1752 (K, holo., B, BR, C, M, S, photo)

Large liana 4–45(–70) m long with white latex; trunk up to 18 cm in diameter; bark dark brown to pale grey, rough; branchlets glabrous. Leaves petiolate; blade obovate, 4–15 cm long, 2–7 cm wide, rounded to shortly and bluntly acuminate, cuneate at the base, with 4–8 pairs of rather straight secondary veins, forming an angle of 45–60° with the midrib; petiole 2–6 mm long, glabrous. Inflorescence terminal, occasionally axillary, 3–11 cm long, dense, many-flowered, puberulous or glabrous; peduncle 5–25 mm long; pedicels 4–7 mm long. Flowers fragrant; sepals (broadly) ovate, 1.2–2 mm long, 1–1.5 mm wide, with 1–3 colleters near the edge; corolla yellow or cream, throat dark red; mature bud head about half as long as tube, ciliate with long curled hairs; tube 4–7 mm long; lobes dolabriform, 6–10 mm long, 4–9 mm wide, truncate at the apex; stamens with apex above or below mouth of corolla tube; style ± 2 mm long; ovules ± 100 in each carpel. Fruit dark brown, 24–75 cm long, 0.3–1.5 cm in diameter; seed up to 23 mm long, coma ± 3 cm long.

UGANDA. Mengo District: Zika Forest, 1 Jul. 1975, *Katende* 2366!; Masaka District: Bale, Lake Nabugabo, 5 May 1969, *Lye* 2785! & ¹/₂ km W of Jubia Forest Station, 13 Aug. 1971, *Katende* 1309!
TANZANIA. Bukoba District: Kantare, 30 Sept. 1935, *Gillman* 302! & Kaigi, May 1935, *Gillman* 281!
DISTR. U 4, T 1; Ivory Coast to Congo (Kinshasa) and Angola
HAB. Moist forest margins; 1100–1200 m
USES. None recorded

SYN. *Wrightia stuhlmannii* K. Schum. in P.O.A. C: 319 (1895). Type: Tanzania, Bukoba District, Bukoba, *Stuhlmann* 3611 (B†, holo., K, lecto., chosen by Leeuwenberg)
Alafia major Stapf in K.B. 1898: 307 (1898). Type: Congo (Kinshasa), Equateur, Wangata, *Dewèvre* 673 (BR, holo.)

4. **A. microstylis** *K. Schum.* in E.J. 23: 230 (1896); Pichon in B.J.B.B. 24: 1881, t. 4/e–f (1954); Kupicha in F.Z. 7 (2): 492 (1985); K.T.S.L.: 475 (1994); Leeuwenberg in K.B. 52 (4): 798, t. 8 (1997). Type: Uganda, sine loc., *Stuhlmann* 1262 (B†, syn.); near Entebbe, *Stuhlmann* 1474 (B†, syn.); neotype: Uganda, Mengo District, L. Victoria, Buvuma Is., *Bagshawe* 620 (K, neo., chosen by Pichon)

Liana 3–15 m high with white latex; branches dark brown with large pale brown lenticels; branchlets glabrous or puberulous. Leaves petiolate; blade (narrowly) elliptic, 3–9 cm long, 1–5 cm wide, bluntly acuminate to obtuse at the apex, cuneate or rounded at the base, glabrous on both sides or puberulous beneath; with 11–14 pairs of rather straight secondary veins at an angle of 40–50° with the midrib; petiole 2–5 mm long, glabrous or puberulous. Inflorescence terminal, few- to many-flowered, 2–5 cm long, glabrous or puberulous; peduncle 1–18 mm long; bracts sepal-like; pedicels 5–15 mm long. Flowers with sepals ovate, up to 2.8 mm long, with 1–2 colleters near the edge inside; corolla white or greenish white; tube often partly reddish, throat red; mature bud with a narrowly ovoid head about twice as long as the tube; tube 4–6.5 mm long; lobes twisted in bud, spoon-like with a shallow concave apex, obliquely obovate or elliptic, 9–14.5 mm long, 3–5.5 mm wide, rounded at apex; stamens inserted 2.2–4 mm from the base; style 2–2.2 mm long. Fruit brown, 40–50 cm long, 0.5 cm in diameter; seed up to 15 mm long; coma 3 cm long.

UGANDA. Ankole District: near Kikagati, Kagera R., Sept. 1947, *Dale* U494!; Toro District: Sempaya Peak, Mar. 1943, *St. Clair-Thompson* F5247!; Mengo District: Kipayo Estate, 1914, *Dümmer* 712!
KENYA. Kilifi District: Kambe Kaya, 3 Mar. 1981, *Hawthorne* 26!; Lamu District: Witu Forest, 16 Nov. 1988, *Robertson & Luke* 5505!; Kwale District: Mrima Hill, 3 Feb. 1989, Mrima-Dzombo Expedition 33!
TANZANIA. Mwanza District: Kibandala, Rubondo Is., Nov. 1954, *Carmichael* 463!; Pangani District: Bushiri Sisal Estate, 14 Mar. 1950, *Faulkner* 549!; Ulanga District: Msolwa Forest, 5 Feb. 1977, *Vollesen* 4393!
DISTR. U 2–4; K 4, 7; T 1, 3, 6, 8; E Congo (Kinshasa), Mozambique
HAB. Forest, riverine forest or bush; 0–1500 m
USES. None recorded

SYN. *A. clusioides* S. Moore in J.L.S. 37: 181 (1905). Type: Uganda, Mengo District, Lake Victoria, Buvuma Is., *Bagshawe* 620 (K, holo.)
  A. swynnertonii S. Moore in J.L.S. 40: 141 (1911). Type: Mozambique, Madanda Forest, *Swynnerton* 1178 (BM, holo., K, iso.)

5. **A. orientalis** *De Wild.*, Not. Apoc. Latic. Fl. Congo 1: 15 (1903); Engl. in N.B.G.B. 3: 84 (1900), nomen; Pichon in B.J.B.B. 24: 206, t. 6/a–b (1954); Kupicha in F.Z. 7, 2: 492 (1985); Leeuwenberg in K.B. 52 (4): 804, t. 10 (1997). Type: Tanzania, Lushoto District, Usambara Mts, Derema, *Scheffler* 69 (B†, holo., BR, neo., chosen by Pichon)

Large liana to 30 m long, with white latex; trunk up to 10 cm in diameter; branches dark brown, with scattered pale brown lenticlels; branchlets glabrous. Leaves petiolate; blade elliptic, 4–14 cm long, 1–6 cm wide, bluntly acuminate to obtuse at the apex, cuneate to rounded at the base, glabrous, with 5–12 pairs of upcurved secondary veins forming an angle of 40–60° with the midrib; petiole 3–8 mm long, glabrous. Inflorescence terminal, 2–6 cm long, 3–6 cm wide, many-flowered, dense, glabrous to puberulous; peduncle 2–7 mm long; pedicels 1–5 mm long. Flowers fragrant; sepals ovate, 2–4 mm long, obtuse, with 1–4 colleters at the edge; corolla orange to yellow with a dark red throat; mature bud with an ovoid or ellipsoid head as long as the tube, ciliate; tube 6–11 mm long; lobes obliquely ovate to obovate, 7–10 mm long, 4.5–8 mm wide, rounded; stamens with apex 0–1 mm below the mouth of the tube, inserted 3.5–5.5 mm from the base; style 2–4 mm long; ovules ± 100 in each carpel. Fruits brown, 20–60 cm long, 0.5–1 cm wide, glabrous; seed up to 16 mm long; coma 3–4 cm long.

UGANDA. Mengo District: Kaazi, Aug. 1953, *Makerere College* 183 & Saza, new H.Q., Dec. 1955, *Dale* 877 & mile 13 Entebbe Road, Nov. 1937, *Chandler* 2026
TANZANIA. Lushoto District: Ngambo, 11 Dec. 1940, *Greenway* 6078 & Kwamkoro, near sawmill, 18 Dec. 1959, *Semsei* 2963! & Monga, Amani, 16 Jan. 1913, *Grote* 4042!
DISTR. U 4; T 3; Congo (Kinshasa), Zimbabwe and Mozambique
HAB. Moist forest and forest margins; 900–1200 m
USES. None recorded

SYN. *A. ugandensis* Pichon in B.J.B.B. 24: 179, t. 4/a–b (1954). Type: Uganda, Mengo District, 21 km on Entebbe road, near Lake Victoria, *Chandler* 2020 (K, holo, BR, P, iso.)

FIG. 32. *ALAFIA SCHUMANNII* — **1**, habit, × ²/₃; **2**, flower, × 2; **3**, sepal inside, × 8; **4**, corolla opened, × 4; **5**, pistil, × 8; **6**, fruit × ¹/₆. 1–5 from *Dawe* 251; 6 from *Louis* 3229. From K.B. 52(4), drawn by E. Catherine.

A. *congolana* Pichon in B.J.B.B. 24: 195, t. 5/a–b (1954). Type: Congo (Kinshasa), Yangambi, Tutuku Is. opposite Isalowe Plateau, *Louis* 13590 (BR, holo., K, P, iso.)
A. *sp.* of T.T.C.L.: 48 (1949)

6. **A. zambesiaca** *Kupicha* in B.J.B.B. 51: 153, t. 1 (1981) & in F.Z. 4, 1: 490, t. 117 (1985); Leeuwenberg in K.B. 52 (4): 814 (1997). Type: Zambia, Kasama District, 5 km E of Kasama, *Robinson* 4041 (K, holo, EA!, M, PRE, SRGH, iso.)

Liana or climbing shrub 2–15 m high, without latex; trunk 1.5 cm in diameter or more; branches dark brown with pale brown lenticels; branchlets glabrous. Leaves petiolate; blade (narrowly) elliptic or ovate, 1.5–6.5 cm long, 0.6–3 cm wide, obtuse to acuminate at the apex, cuneate or rounded at the base, glabrous, with 5–8 pairs of rather straight secondary veins forming an angle of 45–55° with the midrib; petiole 1–3 mm long, glabrous. Inflorescence terminal, 1.5–2.5 cm long, few-flowered, rather lax, glabrous or puberulous; peduncle 2–12 mm long; pedicels 2–10 mm long. Flowers fragrant; sepals (narrowly) ovate, 2–2.6 mm long, 0.8–1 mm wide, acute or acuminate at the apex, with one colleter near the edge; corolla white or cream; mature bud with a narrowly oblong or ovoid head about 2/3 of the bud length, ciliate; tube 6–8.7 mm long; lobes obliquely and narrowly elliptic or oblong, 9–9.3 mm long, 2–3.3 mm wide, rounded, ciliate; stamens with apex above and below mouth of the corolla tube, inserted 3.5–4.5 mm from the base; style 3–4 mm long; ovules ± 50 in each carpel. Fruit brown, 15–40 cm long, 0.3–0.5 cm in diameter; seed 9–15 mm long, 1.5–2.5 mm wide; coma 2.5–3.5 cm long.

TANZANIA. Ufipa District: near Kasanga, 26 Oct. 1933, *Michelmore* 697! & Sumbawanga, Kasanga Escarpment, 24 Nov. 1959, *Richards* 11811! & Kawa R. Falls, Dec. 1956, *Richards* 7408
DISTR. T 4; Congo (Kinshasa), Zambia, Zimbabwe
HAB. Dense bushland or riverine thicket; 900–1500 m
USES. None recorded

7. **A. schumannii** *Stapf* in F.T.A. 4, 1: 197 (1902); Pichon in B.J.B.B. 24: 209, t. 6/c–d (1954); Huber in F.W.T.A. ed. 2, 2: 68 (1963); Leeuwenberg in K.B. 52 (4): 808, t. 12 (1997). Type: Cameroun, Bipindi, *Zenker* 1662 (K, holo., BM, BP, BR, E, G, HBG, L, M, NY, P, S, W, WAG, Z, iso.)

Liana to 40 m long, with little or no white latex; trunk to 20 cm in diameter; branchlets glabrous. Leaves glabrous, blade elliptic, 5–15 cm long, 2–6 cm wide, apex apiculate with acute acumen, base rounded, with 5–8 pairs of upcurved secondary veins forming an angle of 45–60° with the midrib; petiole 5–20 mm long. Inflorescence terminal, many-flowered, 3–6 cm long, dense, glabrous or puberulous; peduncle 3–17 mm long; pedicels 2–7 mm long. Flowers fragrant; sepals ovate, 2.5–3.5 mm long, 1.5–2.5 mm wide, obtuse, with 1–5 colleters; corolla white; mature bud with a narrowly ovoid or ovoid head about half of the bud length, ciliate; tube 8–12 mm long; lobes obliquely suborbicular to ovate or obovate, 6.5–13 mm long, 5.5–11 mm wide, rounded, ciliate; stamens with apex 0–2.5 mm below mouth of the corolla tube, inserted 3.5–5.5 mm from the base; style 3.5–4.5 mm long; ovules ± 120 in each carpel. Fruit dark brown, 40–110 cm long, 0.5–1 cm in diameter; seed 18 mm long, 2–4 mm wide; coma 3–5 cm long. Fig. 32 (p. 97).

UGANDA. Bunyoro District: Budongo Forest, Dec. 1934, *Eggeling* 1476; Ankole District: Kalinzu Forest, June 1938, *Eggeling* 3663
DISTR. U 2; W Africa from Sierra Leone to Angola and Uganda
HAB. Rain-forest; ?1200–1500 m
USES. None recorded

SYN. *Holalafia schumannii* (Stapf) Woodson in Ann. Missouri Bot. Gard. 37: 406 (1950)
    Alafia bequaertii De Wild., Pl. Bequaert. 1: 407 (1922), as *bequaerti*. Type: Congo (Kinshasa), Beni, *Bequaert* 3307 (BR, holo.)

29. MOTANDRA

A. DC., Prodr. 8: 423 (1844); de Kruif in Meded. Landbouwhogeschool 83 (7):
3–17 (1983)

Climbing shrubs or lianas with white latex. Leaves (sub-)opposite, colleters absent
but glands present near base of petiole; blade with domatia in some vein axils.
Inflorescence terminal, thyrsoid, many-flowered; bracts sepal-like or leafy. Flowers 5-
merous. Sepals imbricate, connate at the base, ciliate. Corolla tube with 5 external
pockets appearing as thickenings inside, from about half the length to the base of the
lobes; tube glabrous inside except for tufts of hairs alternating with the stamens;
lobes in bud contorted, overlapping to the right, slightly ciliate; stamens included,
connivent into a cone; anthers with penicellate hairs at the apex, at both tips with a
subglobose thickening up to 0.1 mm in diameter, between the tails on the connective
an orbicular patch, the retinacle, with long hairs along the margin adhering to the
pistil head; ovary semi-inferior, carpels two, connate at the extreme base, entirely or
apically densely pubescent; disk present, adnate to the ovary at the base, composed
of a ring and 5 lobes, alternating with the stamens; style obconical to cylindrical; pistil
head composed of a slightly 5-winged upper and a cylindrical lower portion to which
5 retinacles adhere. Fruits pendulous, of two woody follicles, connate at the extreme
base, dehiscent adaxially. Seeds attached to the adaxial side of the fruit, flat, elliptic
to ovate; coma directed towards the apex of fruit.

Three species in tropical Africa.

M. guineensis (*Thonn.*) *A. DC.* in Prodr. 8: 423 (1844); Stapf in F.T.A. 4, 1: 224 (1902);
Huber in F.W.T.A. ed. 2, 2: 80 (1963); de Kruif in Meded. Landbouwhogeschool 83 (7):
5, t. 1, photo 1 (1983). Type: Ghana, Akwapim, *Thonning* 262 (C, holo., P-JU, iso.)

Climbing shrub or liana 0.5–40 m high; trunk 1–10 cm in diameter; bark brown,
smooth, later longitudinally fissured; branchlets densely dark brown-pubescent.
Leaves petiolate; blade elliptic to oblong-obovate, 3.5–14 cm long, 1.4–4.6 cm wide,
sometimes ciliate with long hairs; domatia pale brown, sometimes absent; petiole
3.5–10(–13) mm long, rusty brown-pubescent, glabrescent, with glands at the base.
Inflorescence rather lax, 2.5–15.5 cm long, 1.5–7(–9) cm wide; peduncle and
branches rusty brown-pubescent, glabrescent; bracts ovate to triangular, inside with
1–3 colleters. Flowers fragrant; sepals triangular, 1–1.9 mm long, 0.3–1.2 mm wide,
with 0–2 colleters per sepal; corolla in mature bud 4.8–11 mm long; tube white to
greenish white, obconical, 2.5–4 mm long, inside with tufts of hairs 0.9–1.9 mm from
the base; lobes white to greenish white, narrowly ovate to obovate, 2.3–7 mm long,
1–2.1 mm wide; stamens inserted at 0.2–0.4 mm from the base of the tube; penicellate
hairs 0.4–0.5 mm long; pistil 2.5–3.1 mm long; style 0.2–0.7 mm long; ovules 25–40 in
each carpel. Fruit dark green with dense rusty brown indumentum, and numerous
wing-like ridges, 4–18 cm long, 1–3.5 cm in diameter; exocarp 1–2 mm thick; seeds
numerous, 11–18.5 mm long, 5–8.5 mm wide; coma 29–78 mm long. Fig. 33 (p. 100).

UGANDA. Bunyoro District: Siba, 4 May 1965, *Karani* 29! & Bujenje, Budongo Forest, 29 Oct.
1971, *Synnott* 725!; Mengo District: Buvuma, near Katamiro, 26 Sept. 1949, *Dawkins* 401!
DISTR. U 2, 4; West and Central Africa from Guinea to Uganda and S to Angola
HAB. Light or secondary deciduous forest or gallery forest and in secondary regrowth, on sand,
clay and rocky outcrops; 0–1200 m
USES. None recorded

SYN. *Echites guineensis* Thonn. in Schum., Beskr. Guin. Pl.: 149 (1827)
    Motandra altissima Stapf in J.L.S. 37: 526 (1906). Lectotype: Uganda, Mengo District,
    Bunjako I., *Dawe* 260 (K, lecto., chosen by de Kruif)

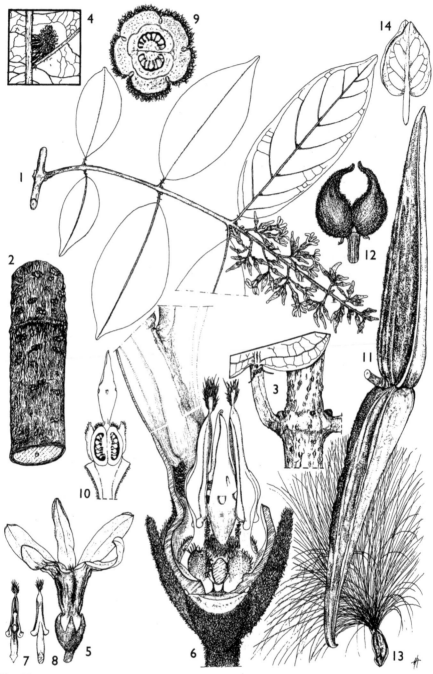

FIG. 33. *MOTANDRA GUINEENSIS* — **1**, flowering branch, × ²/₃; **2**, part of branch, × ²/₃; **3**, leaf base with petiole, × 4; **4**, domatium, × 4; **5**, flower, × 4; **6**, section of flower, × 20; **7**, stamen adaxial side, × 6; **8**, stamen abaxial side, × 6; **9**, TS ovary, × 10; **10**, LS ovary with pistil, × 10; **11**, fruit, × ²/₃; **12**, immature fruit, × 2; **13**, seed, × ²/₃; **14**, embryo, × 2. 1 from *Binuyo* 41203; 2 from *de Kruif* 665; 3 from *de Kruif* 658; 4, 11 from *de Kruif* 722; 5–10 from *Leeuwenberg* 11974; 12–14 from *Gossweiler* 5223B. From Meded. Landbouwhogeschool 83(7), drawn by Y.F. Tan, reproduced by permission.

30. BAISSEA

A.DC., Prodr. 8: 424 (1844); de Kruif in F.Z. 7 (2): 495 (1985); van Dilst in B.J.B.B. 64: 89–178 (1995).

Zygodia Benth. in G.P. 2: 716 (1879)

Climbing shrubs, lianas or rhizomatous creepers (not in our area) with white latex; tendrils, axillary glands and stipules absent. Leaves opposite, petiolate; petiole with colleters above, near base to halfway up; domatia usually in some vein axils. Inflorescences axillary and terminal, few to many-flowered cymes. Flowers 5-merous, fragrant or not; sepals imbricate, with 0–10 colleters at the base inside. Corolla tube with dense tufts of hairs alternating with the stamens inside, usually with scales or ridges above the insertion of the stamens; lobes contorted in bud, overlapping to the right; stamens inserted in lower part of corolla tube; anthers connivent into a cone over the pistil head, coherent with the pistil head by a circular patch, the retinacle; ovary semi-inferior, carpels two, connate at the extreme base, disk present, adnate to the ovary base; pistil head with a nearly cylindrical basal part, adhering to the anthers, a slightly 5-winged central part and a bilobed stigmoid apex. Fruits of 2 pendulous, follicular mericarps, connate at the extreme base and sometimes at the apex, usually slightly constricted between the seeds; seeds with a deciduous apical coma directed towards the apex of the mericarp.

18 species, restricted to continental Africa.

1. Corolla lobes 2–3 mm long; coastal species · · · · · · · · · · · ·   4. *B. myrtifolia*
   Corolla lobes > 4 mm long; inland species · 2
2. Corolla tube pubescent above scales inside · · · · · · · · · · ·   5. *B. viridiflora*
   Corolla tube glabrous above scales inside · 3
3. Leaf with a rounded or obtuse base; corolla tube mostly
   cylindrical, glabrous outside · · · · · · · · · · · · · · · · · · ·   3. *B. major*
   Leaf with a cuneate base; corolla tube funnel-shaped or
   campanulate, pubescent outside · 3
4. Branches minutely pubescent or glabrous; secondary veins
   5–10 mm apart; corolla lobes 10–12 mm long · · · · · · · · ·   1. *B. leonensis*
   Branches densely dark brown-pubescent; secondary veins
   15–20 mm apart; corolla lobes 4.8–8 mm long · · · · · · · ·   2. *B. leontonori*

1. **B. leonensis** *Benth.* in Hook., Niger Fl.: 452 (1849); Stapf in F.T.A. 4, 1: 213, 611 (1902); Huber in F.W.T.A. ed. 2, 2: 78 (1963); van Dilst in B.J.B.B. 64: 112, t. 11 (1995). Lectotype: Sierra Leone, sine loc., *Vogel* 119 (K, lecto.; BR, MO, NY, WAG, photo K, chosen by van Dilst)

Liana up to 40 m high or scandent shrub 3–8 m high; trunk up to 15 cm in diameter; bark red-brown, rough or smooth; branchlets minutely pubescent or glabrescent. Leaves petiolate; blade narrowly elliptic-oblong, rarely obovate, 3–8.5 cm long, 1.5–3 cm wide, acute to apiculate at the apex, cuneate, sometimes acute at the base, glabrous or puberulous to glabrescent; secondary veins in 6–17 pairs; domatia present, often red; petiole 3–20 mm long, minutely pubescent or glabrescent, with or without colleters above. Inflorescence axillary and terminal, few to many-flowered, dark pubescent in all parts; peduncles 2–29 mm long; pedicels 2–6.5 mm long. Flowers fragrant; sepals deltoid or narrowly ovate, 1–3.4 mm long, ciliate, with 3–6 colleters at the base; corolla tube white or yellow, sometimes purple-shaded outside, reddish-striped inside, funnel-shaped, 4–5.5 mm long, scales distinct or not, tufts of hairs at 0.3–0.7 mm from the base; corolla lobes white with a purple tinge or entirely pink, purple, red or yellow, narrowly ovate or oblong, 10–12 mm long,

1–2.5 mm wide, obtuse at the apex; ovary cylindrical or subglobose; style up to 1 mm long, pubescent. Fruits dark red brown or grey brown, ribbed or smooth, sparsely pubescent, mericarps 15–105 cm long, up to 14 mm in diameter; seeds black, up to 32 mm long, coma 23–34 mm long.

UGANDA. Mengo District: near Entebbe, Feb. 1932, *Eggeling* 421! & Nyenga, *Bagshawe* 584 & Lutembe Bay, 14 km on Entebbe road, 28 Mar. 1959, *Lind* 2384!
DISTR. **U** 4; from Senegal to Uganda in the E and to Angola in the S
HAB. Moist forest and forest margins; 1100–1200 m
USES. None recorded

SYN. *B. alborosea* Gilg & Stapf in Z.A.E. 2: 540 (1913). Type: Congo (Kinshasa), Beni Urumu, Pojo, *Mildbraed* 2826 (B†, holo., K, lecto., chosen by van Dilst)
   B. giorgii De Wild. in B.J.B.B. 5: 108 (1915). Type: Congo (Kinshasa), Mobwasa, *De Giorgi* 794 (BR, holo.)
   B. likimiensis De Wild., Pl. Bequaert. 5: 419 (1932). Type: Congo (Kinshasa), Likimi, *Malchair* 501 (BR, holo.)

2. **B. leontonori** *van Dilst* in B.J.B.B. 64: 120, t. 13 (1995). Type: Burundi, Kwitaba, *Reekmans* 6531 (WAG, holo., AAU, BR, K, LG, iso.)

Shrub or liana up to 8 m high; branches greyish dark brown, pubescent; branchlets densely pubescent. Leaves petiolate; blade narrowly elliptic to oblong, 3.4–11.5 cm long, 1.5–3 cm wide, acuminate or apiculate at the apex, rarely obtuse, cuneate at the base, glabrous above, pubescent or puberulous beneath, with 7–12 pairs of secondary veins, 2nd–7th vein conspicuously arcuate; domatia present; petiole 1.5–7 mm long, pubescent, with colleters half way up and sometimes with 1–2 in the axils. Inflorescence axillary and terminal, few-flowered cymes, indumentum dark red brown or dark grey; peduncles 1.1–10 mm long; pedicels 2–5.5 mm long. Flowers fragrant; sepals ovate or triangular, up to 3 mm long, with 5–9 colleters at the base; corolla white or creamy, sometimes with black spots, inside red-striped; tube funnel-shaped, 4.8–6.5 mm long, (dark) grey-pubescent outside, with distinct scales 1.5–2 mm from the base; lobes ovate or strap-shaped, 4.8–8 mm long, 2–5 mm wide, obtuse or subacute at the apex, variably grey-pubescent outside, glabrous inside, variably ciliate; ovary subglobose, entirely black-pubescent or only at the apex; style 0.6–1.7 mm long, minutely pubescent. Fruit follicles rusty brown, minutely pubescent, 20–30 (50) cm long, 0.4–0.5 cm in diameter; seed black or dark red-brown, up to 18 mm long, coma pale red, 17–25 mm long.

UGANDA. Masaka District: N of Muwawu Forest, Kalungu, 13 Jun. 1971, *Katende* 980! & 2 km NE of Nkoma, 13 Jun. 1971, *Lye & Katende* 6234!
DISTR. **U** 4; Eastern Congo (Kinshasa), Burundi
HAB. Forest or ravine bushland; 1150–1400 m
USES. None recorded

3. **B. major** (*Stapf*) *Hiern* in Cat. Afr. Pl. Welw. 1: 675 (1898); Stapf in F.T.A. 4, 1: 210 (1902); van Dilst in B.J.B.B. 64: 128, t. 17 (1995). Type: Angola, Pungo Andongo, *Welwitsch* 5966 (K, holo., BM, COI, LISC, iso.)

Scandent shrub up to 5 m high or liana up to 15 m high; trunk up to 8 cm in diameter; branches dark red-brown, smooth; branchlets minutely pubescent or puberulous. Leaves petiolate; lamina obovate or elliptic, 3.4–16 cm long, 1.3–6.2 cm wide, acuminate or apiculate, rarely acute to obtuse at the apex, obtuse to rounded at the base, pubescent on midrib and secondary veins above, red-pubescent beneath; secondary veins in 7–12 pairs, domatia present; petiole 3–10 mm long, pubescent, with colleters above. Inflorescence axillary and terminal, few- to many-flowered cymes, pubescent; peduncle red, 7–28 mm long; pedicles dark red, 1.5–4 mm long. Flowers fragrant; sepals dark red, ovate or oblong, 1.6–3.9 mm long, variably

Fig. 34. *BAISSEA MAJOR* — **1**, flowering branch, × ²/₃; **2**, part of leaf, × 1; **3**, leaf with domatium beneath, × 10; **4**, petiole and leaf base, × 1; **5**, petiole and leaf base with colleters, × 5; **6**, flower, × 5; **7**, opened flower, × 10; **8**, sepal with colleters, × 9; **9**, pistil, × 28; **10**, stamen side view, × 28; **11**, stamen inside, × 30; **12**, fruit, × ²/₃; **13**, seed × 1. 1, 3, 5–11 from *Hart* 501; 2 from *Hart* 404; 12–13 from *Tisserant* 2099. From B.J.B.B. 64, drawn by H. de Vries, reproduced by permission.

pubescent outside, ciliate, with colleters within; corolla greenish in bud, white at anthesis; tube 3.5–4.9 mm long; scales distinct, 1.2–1.5 mm from the base; corolla lobes ovate or strap-shaped, 6–24 mm long, obtuse or acute at the apex, glabrous to puberulous outside, glabrous inside; disk 0.1–1.3 mm long; ovary subglobose or cylindrical variably pubescent; style 0.7–1 mm long, pubescent. Fruit follicles dark red-brown, 38–75 cm long, 0.3–0.5 mm wide, puberulous or glabrescent; seeds black, 12–21 mm long, coma pale red, 18–26 mm long. Fig. 34 (p. 103).

UGANDA. Masaka District: 3 km W of Katera, 4 Oct. 1953, *Drummond & Hemsley* 4618!; Mengo District: Mukono, *Dümmer* 2444
KENYA. North Kavirondo District: Kakamega Forest, Nov. 1934, *Dale* 3385! & ± 1923, *Battiscombe* K1212!
TANZANIA. Bukoba District: Kagera, Nov. 1994, *Congdon* 374
DISTR. **U** 4; **K** 5; **T** 1; Cameroon, Central African Republic, Congo (Kinshasa) and Angola
HAB. Moist forest and swamp forest; 1150–1600 m
USES. None recorded

SYN. *B. angolensis* Stapf var. *major* Stapf in K.B. 1894: 126 (1894)
     B. multiflora sensu auctt. e.g. U.K.W.F. ed. 2: 171 (1994); K.T.S.L.: 477 (1994), *non* A.DC.

4. **B. myrtifolia** (*Benth.*) *Pichon* in Bull. Mus. Hist. Nat. Paris, ser. 2 (20): 196 (1948); de Kruif in F.Z. 7 (2): 497 (1985); K.T.S.L.: 477, ill., map (1994); van Dilst in B.J.B.B. 64: 136, t. 21 (1995). Type: Tanzania, Bagamoyo District, Zaramu, *Kirk* s.n. (K, holo.)

Scandent shrub up to 5 m high or liana up to 20 m high; trunk up to 2 cm in diameter; branches slender, grey to red-brown; branchlets densely pubescent. Leaves petiolate; blade obovate to oblong, 2–8.1 cm long, 1.1–3.7 cm wide, acute to apiculate, sometimes obtuse at the apex, cuneate to rounded at the base, pubescent or glabrous; secondary veins in 6–12 pairs, reddish domatia present; petiole 1–5.6 mm long, pubescent, with colleters. Inflorescences axillary and sometimes also terminal, pubescent; peduncles 0.5–16 mm long; pedicels 0.7–2.2 mm long. Flowers fragrant; sepals narrowly ovate or oblong, 1.7–2.8(–4.1) mm long, (0.4–)0.6–1 mm wide, pubescent outside, with colleters within; corolla bright yellow, with pink throat fading orange; tube cylindrical, urceolate or funnel-shaped, 1.8–4 mm long, glabrous or pubescent outside; scales united into a recurved ridge, 1–2.3 mm from the base of the tube; lobes ovate or sometimes oblong, 2–3 mm long, 0.9–1.4 mm wide, obtuse or subacute at the apex, glabrous or pubescent outside, pubescent inside; ovary cylindrical or subglobose, pubescent at the apex; style 0.2–0.9 mm long, minutely pubescent or papillose; disk 0.1–0.3 mm high. Fruit follicles red-brown, 20–30 cm long, 0.2–0.5 cm in diameter, glabrous, brown-pubescent when young; seeds black 9–18 mm long, 1.6–1.8 mm wide; coma 18–38 mm long.

KENYA. Kwale District: Buda Forest, Nov. 1936, *Dale* 3580!; Malindi District: Mangea Hill, 28 Dec. 1988, *Luke* 1602; Malindi District: Arabuko Sokoke Forest, *Musyoki & Hansen* 1004!
TANZANIA. Tanga District: Pongwe, Mawasi, 12 Apr. 1968, *Faulkner* 4103!; Uzaramo District: near Kibaha, 35 km W of Dar, 22 Apr. 1970, *Flock* 56/J!; Lindi District: Mingoyo, 23 Mar. 1943, *Gillman* 1343!
DISTR. **K** 7; **T** 3, 4, 6, 8; **Z**; not known elsewhere
HAB. Evergreen forest, thicket or coastal bush; 0–450(–900) m
USES. None recorded

SYN. *Zygodia myrtifolia* Benth. in Hook., Ic. Pl. new ser. 2: t. 1184 (1876); Stapf in F.T.A. 4, 1: 218 (1902); T.T.C.L.: 57 (1949)
     Oncinotis melanocephala K. Schum. in Phys. Abh. Kön. Akad. Wiss. Berlin 1: 34 (1894) & E. & P. Pf. 4 (2): 179 (1894) & P.O.A. C: 319 (1895). Type: Tanzania, Lushoto District, Usambara Mts, Misosue, Buruka R., *Holst* 2217 (B†, holo., K, lecto., COI, HBG, iso., chosen by van Dilst)

Zygodia kidengensis K. Schum. in P.O.A. C: 318 (1895) & E. & P. Pf. 4 (2): 164 (1895); T.T.C.L.: 57 (1949). Type: Tanzania, Bagamoyo District, Zaramu, Kedenge, *Stuhlmann* 6344 (B†, holo., K, lecto., P, iso., chosen by van Dilst)

Z. melanocephala (K. Schum.) Stapf in F.T.A. 4, 1: 219 (1902); T.T.C.L.: 57 (1949)

Baissea melanocephala (K. Schum.) Pichon in Bull. Mus. Nat. Hist. Nat. Paris, ser. 2 (20): 196 (1948)

5. **B. viridiflora** (*K. Schum.*) *de Kruif* in Meded. Landbouwhogeschool 83 (7): 16 (1983) & in F.Z. 7 (2): 497 (1985); van Dilst in B.J.B.B. 64: 151, t. 29 (1995). Type: Tanzania, Lushoto District, Usambara Mts, Nguelo, *Scheffler* 28 (B, holo., E, K, P, iso.)

Liana up to 15 m high or scandent shrub up to 8 m high; branches pale grey to brown; branchlets pubescent or glabrescent. Leaves petiolate; blade obovate to elliptic, 4.2–11 cm long, 1.7–4.5 cm wide, obtuse to cuspidate at the apex, mostly acuminate, cuneate at the base, glabrous above, glabrous or glabrescent beneath; secondary veins in 5–10 pairs, domatia red and mostly palisade-like; petiole 2–10 mm long, glabrous or minutely pubescent, with colleters above. Inflorescences axillary and/or terminal, pubescent; peduncle 3–61 mm long; pedicels 0.9–4.6 mm long. Flowers fragrant; sepals ovate or narrowly ovate, 0.5–2.7 mm long, obtuse to subacute at the apex, minutely pubescent outside, ciliate, with colleters within; corolla white or yellow; tube 1.3–2.9 mm long, glabrous or minutely pubescent outside, scales indistinct, with tufts of hairs inside; lobes ovate to strap-shaped, 5–10 mm long, 0.8–1.8 mm wide, obtuse to subacute at the apex; disk 0.1–0.2 mm high, pistil 1.1–2.1 mm high; ovary cylindrical or subglobose, glabrous or pubescent at the apex; style 0.2–0.6 mm long. Fruits follicles brown-pubescent or glabrescent, 6–30 cm long, 2–5 mm in diameter; seeds 12–13 mm long; coma 14–16 mm long.

UGANDA. Kigezi District: Amahenge, Kinkizi, Kigezi Forest, *Purseglove* 2019!

TANZANIA. Lushoto District: East Usambaras, Monga Forest, 27 Jul. 1974, *Baagøe* 99! & between Monga and Ndola, 24 May 1950, *Verdcourt* 214!; Morogoro District: Turiani, Nov. 1953, *Semsei* 1479!

DISTR. **U** 2; **T** 3, 6; Eastern Congo (Kinshasa) and Malawi

HAB. Moist forest or gallery forest; 750–1500 m

USES. None recorded

SYN. *Motandra viridiflora* K. Schum. in E.J. 33: 319 (1903); Stapf in F.T.A. 4, 1: 613 (1904); T.T.C.L.: 53 (1949)

31. ONCINOTIS

Benth. in Niger Fl.: 451–452 (1849); G. P. 2: 718 (1876); Pichon in B.J.B.B. 24: 9–36 (1954); de Kruif in Agric. Univ. Wageningen Papers 85 (2): 8–45 (1985)

Climbing shrubs or lianas; white latex present; branchlets (sub-)opposite, terete. Leaves (sub-)opposite; petiole with glands on the adaxial side; domatia in some vein axils, consisting of pits, often with a ciliate margin or of a tuft of hairs. Inflorescences terminal or axillary, thyrsoid, many-flowered; bracts deciduous. Flowers 5-merous, fragrant. Sepals imbricate, ciliate. Corolla tube shortly pubescent outside, inside with tufts of stiff hairs, alternating with the stamens, at the mouth with 5 alternipetalous corona scales; lobes overlapping to the right, entire, recurved; stamens included, connivent into a cone; filaments short; anthers partly fertile, below the fertile part inside a nearly oblong patch, the retinacle, with minute hairs along the margin adhering to the pistil head; ovary semi-inferior; carpels two, connate at the extreme base, densely pubescent; disk present, adnate to the ovary at the extreme base, composed of a ring and five lobes; pistil head of a slightly 5-winged upper and a cylindrical lower portion to which the 5 retinacles adhere and a bifid stigmoid. Fruits pendulous, of two follicles connate at the extreme base, dehiscent along an adaxial line; exocarp woody; endocarp stiff, thinly pergamentaceous. Seeds attached to the adaxial side of the fruit, with an apical coma directed towards the apex of the fruit.

Six species in Tropical Africa and one in Madagascar.

1. Branchlets and leaves densely covered with variously branched hairs; domatia absent or inconspicuous, consisting only of tufts of hairs · 2. *O. hirta*
   Branchlets and leaves usually glabrous, rarely puberulous or pubescent but then with simple hairs; domatia usually present · 2
2. Leaves with inconspicuous tertiary venation beneath; leaf blade obovate, often emarginate at the apex, rarely shortly and obtusely acuminate; pistil 2–2.8 mm long · · · · · · · · · 3. *O. pontyi*
   Leaves with conspicuous tertiary venation beneath; leaf blade elliptic, ovate to obovate and distinctly acuminate at the apex; pistil 2.8–3.5 mm long · 3
3. Branchlets glabrous, rarely puberulous; leaf blade elliptic to ovate; secondary veins forming an angle of 50–75° with the midrib; follicles longitudinally winged when mature · · · · 1. *O. glabrata*
   Branchlets puberulous or pubescent; leaf blade obovate; secondary veins strongly curved, forming an angle of 10–40° with the midrib; follicles narrowly cylindrical, not winged · 4. *O. tenuiloba*

1. **O. glabrata** (*Baill.*) *Hiern* in Cat. Afr. Pl. Welwitsch 1(3): 674 (1898); Stapf in F.T.A. 4, 1: 222 (1902); Huber in F.W.T.A. ed. 2, 2: 80 (1963); de Kruif in Agric. Univ. Wageningen Papers 85 (2): 13, t. 1 (1985). Type: Angola, Cuanza Norte, Ambaca road, *Welwitsch* 5957 (P, holo., BM, COI, G, K, LISU, MO, iso.)

Climbing shrub or liana 1.2–40 m high; trunk 1.5–12 cm in diameter; bark greyish brown, with large pale brown lenticels; branchlets brownish green to grey, glabrous or rarely puberulous. Leaves petiolate; blade elliptic to ovate (to obovate), 3.9–13 cm long, 1.5–5 cm wide, acuminate (or rounded or emarginate) at the apex, cuneate at the base, glabrous (rarely puberulous on midrib and veins), glossy, dark to medium green above, with 5–12 pairs of secondary veins forming an angle of 50–75°, tertiary venation reticulate to faintly scalariform; domatia of pits, sometimes absent, sometimes with a ciliate margin; petiole 7–24 mm long, glabrous, rarely puberulous, with 1–3 clusters of 2–3 glands on the adaxial side. Inflorescence (2.5–)4–12.5 cm long, rusty brown-pubescent and glabrescent in all parts; bracts ovate to triangular, 1.2–1.7 mm long; pedicels 2–5 mm long. Flowers with sepals ovate, 1.5–3.6 mm long, rusty brown-pubescent; corolla yellow; tube urceolate, 2.5–4.7 mm long; lobes triangular, 2.3–5.8 mm long, 0.9–2.5 mm wide, glabrous or puberulous outside; corona later turning white, 0.5–0.9 mm long, pubescent at the base; pistil 2.9–3.5 mm long; style 0.1–0.2 mm long; ovules 60–120. Fruits fusiform, 9.8–30 cm long, 1.2–6.1 cm wide, exocarp 1–8 mm thick, very hard, longitudinally winged at maturity, wing undulate, 0.1–3 mm wide, puberulous to glabrescent; seeds numerous, 5–27 × 2–6.5 mm; coma 10–75 mm long.

UGANDA. Mengo District: 1.5 km NE of Entebbe town, below Katabi Hill, 18 Oct. 1950, *Dawkins* 659! & Kipayo Estate, May 1914, *Dümmer* 809! & near Mpoma in Kifu Forest, Apr. 1969, *Lye et al.* 2588
TANZANIA. Mpanda District: Kungwe, Mahali Peninsula, near head of Ntali R., 9 Sept. 1959, *Harley* 9569!
DISTR. **U** 4; **T** 4; tropical Africa from Guinea to Central African Republic and S from to Congo (Kinshasa), Burundi and Angola
HAB. Moist forest, riverine forest, secondary bushland; 1100–1800 m
USES. None recorded

SYN. *Motandra glabrata* Baill. in Bull. Soc. Linn. Paris 1: 760 (1888)
   Oncinotis jespersenii De Wild., Not. Pl. Util. Congo 2 (2): 256 (1908), as *jesperseni*. Type:
   Congo (Kinshasa), Equateur, near Mondombe, *Jespersen* 12 (BR, holo.)

2. **O. hirta** *Oliv.* in Hook. Ic. Pl. 13: 25, t. 1232 (1877); Stapf in F.T.A. 4, 1: 223 (1902);
Pichon in B.J.B.B. 24: 32, t. 1 (1954); Kupicha in F.Z. 7 (1): 492 (1985); de Kruif in
Agric. Univ. Wageningen Papers 85 (2): 21, t. 3 (1985). Type: Congo (Brazzaville),
Loango near Makongo [Makunga], *Soyaux* 147 (K, holo., BP, K, M, W, iso.)

Climbing shrub or liana 3–25 m high; trunk 1–6 cm in diameter; bark greyish
brown, ± rough; branches greyish brown, with many large pale brown lenticels;
branchlets densely reddish brown-pubescent. Leaves petiolate; blade elliptic-oblong
to obovate, 4–13.5 cm long, 2–6.8 cm wide, acuminate, exceptionally emarginate at
the apex, rounded or cuneate at the base, with variously branched hairs; domatia
absent or very inconspicuous; petiole 4–12(–22) mm long, densely covered with
variously branched hairs, with 1–15 glands on the adaxial side and 2(–8) glands at
the base of the blade. Inflorescence 3–11 cm long, with very dense rusty brown
tomentum in all parts; bracts up to 1.4 mm long; pedicels 1–2.5 mm long. Flowers
with sepals ovate, 1.2–2.6 mm long, with dense rusty brown tomentum; corolla tube
yellow to greenish yellow, urceolate, 2.7–3 mm long; lobes yellow to greenish yellow,
ovate to triangular, 2.3–2.8 mm long, 0.8–1.2 mm wide, densely pubescent outside;
corona yellowish later turning white, 0.2–0.5 mm long, glabrous, puberulous at the
base; stamens inserted 0.6–0.8 mm from the base of the tube; pistil 1.9–2.5 mm long;
style 0.1–0.2 mm long; ovules 15–30 in each carpel. Fruits narrowly cylindrical,
6.3–24 cm long, 0.4–1 cm in diameter, covered with dense, up to 2.5 mm long
deciduous rusty brown or grey tomentum; seeds up to 25 in each follicle; seed
9.5–19 mm long; coma 21–45 mm long.

UGANDA. Mengo District: Kipayo, Aug. 1915, Dummer 1392 & Kitabe, near Entebbe, Jun.–Jul.
   1935, *Chandler* 1270! & 5–10 km NW of Kisubi, W Buganda, 9 Nov. 1972, *Katende* 1729!
DISTR. U 4; Cameroon, Central African Republic, Gabon, Congo (Brazzaville), Congo (Kinshasa)
   and Angola
HAB. Moist forest or forest margin; 1100–1200 m
USES. None recorded

3. **O. pontyi** *Dubard* in Not. Syst., Paris 2: 201 (1911); Pichon in B.J.B.B. 24: 26, t.
1 (1954); de Kruif in Agric. Univ. Wageningen Papers 85 (2): 29, t. 5, phot. 1 (1985).
Type: Ghana, Amnafo, *Giraud* s.n. (P, holo.)

Liana 5–50 m high; trunk 1–15 cm in diameter; branches greyish black-brown, with
many orange to grey-brown lenticels; branchlets brownish grey to pale grey, smooth,
with many orange to grey-brown lenticels, glabrous or sparsely pubesent. Leaves
petiolate; blade obovate, 4–18 cm long, 1.4–10 cm wide, acuminate, often
emarginate at the apex, cuneate at the base, glabrous, rarely puberulous on midrib,
dark to medium green and glossy above, dull and distinctly paler beneath, with 3–10
pairs of secondary veins; tertiary veins inconspicuous; domatia of pits with ciliate
margin, sometimes absent; petiole 7–22 mm long, glabrous, rarely slightly pubescent,
glabrescent, with 1–3 pairs of glands on the adaxial side. Inflorescence 3–11.5 cm
long, rusty brown-pubescent; bracts ovate to triangular, up to 1.6 mm long; pedicels
2–3.5 mm long. Flowers with sepals broadly ovate to triangular, 0.8–2.1 mm long,
0.6–1.2 mm wide, rusty brown-pubescent; corolla yellowish or whitish green; tube
cylindrical, 2.2–3.8 mm long; lobes narrowly ovate, 1.7–5.8 mm long, 1–1.7 mm wide,
outside puberulous to glabrous; corona later white, 0.4–0.8 mm long, pubescent or
glabrous at the base; stamens inserted 0.6–1 mm from the base of the tube; style
0.1–0.2 mm long; ovules 30–50. Fruits narrowly cylindrical, 12–29 cm long, 0.4–2.6 cm

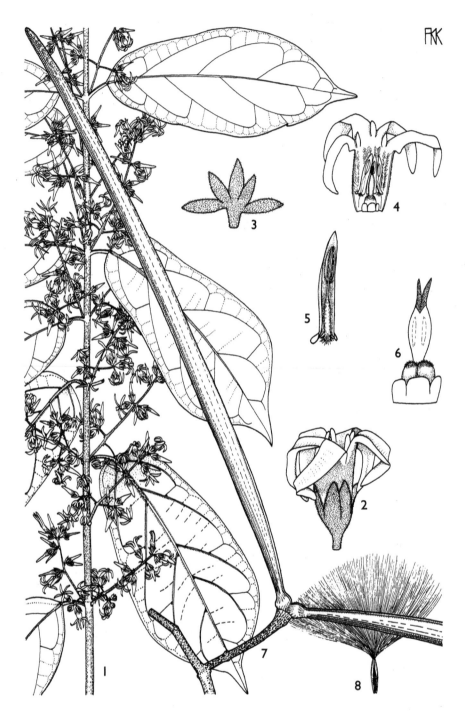

FiG. 35. *ONCINOTIS TENUILOBA* — **1**, habit × ²/₃; **2**, flower, × 4; **3**, calyx opened out, × 4; **4**, part of corolla opened, × 4; **5**, stamen, × 12; **6**, gynoecium with disk scales, × 12; **7**, fruit × ²/₃; **8**, seed × ²/₃. 1–6 from *Simão* 485; 7 from *Drummond* 10187; 8 from *Wild* 2175. From F.Z., drawn by F.K. Kupicha.

wide, very hard greyish to rusty brown pubescent, glabrescent, exocarp 2 mm thick; seeds up to 40 in each follicle, 16–27 mm long; coma 45–93 mm long.

UGANDA. Bunyoro District: Budongo Forest, June 1936, *Eggeling* 3035; Mengo District: Mabira Forest, Nov. 1919, *Dümmer* 4348 & Katambwa R., 8 Aug. 1974, *Katende* 2230
DISTR. U 2, 4; Guinea, Ivory Coast, Ghana, Nigeria, Gabon, Cameroon, Central African Republic, Congo (Kinshasa) and Sudan
HAB. Moist forest; 1100–1200 m
USES. None recorded

SYN. *O. obovata* De Wild. in B.J.B.B. 5: 103 (1915) & 7: 22 (1920). Lectotype: Congo (Kinshasa), Kasoui, Sankuru Forest, March 1907, *Luja* s.n. (BR, holo., G, iso., chosen by de Kruif)
    O. pontyi Dubard var. *breviloba* Pichon in B.J.B.B. 24: 28, t. 2/d–e, 17, t. 1 (1954). Lectotype: Congo (Kinshasa), km 14 on Yangambi–Weko road, *Louis* 16055 (P, holo., BR, C, K, iso., chosen by de Kruif)

4. **O. tenuiloba** *Stapf* in K.B. 1898: 307 (1898); Kupicha in F.Z. 7 (2): 493, t. 118 (1985); de Kruif in Agric. Univ. Wageningen Papers 85 (2): 33, t. 6 (1985); K.T.S.L.: 482, ill. (1994). Type: Congo (Kinshasa), Equateur, near Likasa [Lukasa], *Dewèvre* 883 (BR, holo.)

Climbing shrub or liana 1.8–30 m high; bark light yellow-orange, thick, corky; inner bark with rope-like fibres; wood creamy; branches pale to dark brown, with many whitish lenticels; branchlets brownish to greyish green, puberulous to pubescent. Leaves petiolate; blade obovate, (1.8–)4.6–14.5 cm long, 1.2–5.5 cm wide, acuminate, rarely emarginate at the apex, base cuneate, glabrous, rarely pubescent, with 2–7 pairs of secondary veins; tertiary venation conspicuous, scalariform; domatia of pits with a dense tuft of hispid hairs along the margin; petiole 5–12 mm long, puberulous or pubescent, with 2 pairs of glands along the adaxial side, one pair near the proximal end of the midrib. Inflorescence 2.5–6.5 cm long, rusty brown-pubescent; bracts up to 3.1 mm long; pedicels 1.2–4 mm long. Flowers with sepals elliptic to triangular, 1.2–3.2 mm long, 0.8–1.5 mm wide, rusty brown pubescent or puberulous; corolla yellow-green, tube barrel-shaped, 2.5–3.8 mm long, lobes 3–7.3 mm long, 0.7–1.5 mm wide, glabrous or puberulous outside; corona 0.4–1 mm long; stamens inserted 0.5–1 mm from the base of the tube; pistil 2.8–3.3 mm long; style 0.1–0.3 mm long; ovules 15–45 in each carpel. Fruits pendulous, narrowly cylindrical, tapering at both ends, 10.5–30 cm long, 0.4–1.3 cm wide, greyish to dark brown-pubescent; seeds up to 35 in each follicle; seed 12–23 mm long, coma 25–58 mm long. Fig. 35 (p. 108).

UGANDA. Toro District: Kanyawara, 1973, *Rudran* 101!; Busoga District: Butembe Bunya, near Kigoma, 23 Nov. 1950, *Wood* 16!; Mengo District: Kasala Forest Reserve, Jan. 1915, *Dümmer* 1416!
KENYA. Nandi District: Kaimosi Forest, 10 Jun. 1989, *Mungai* 150/89!; Kwale District: Jego, 27 May 1990, *Luke & Robertson*, 2325!; Tana River District: Tana R. Primate Reserve, 12 Mar. 1990, *Luke et al.* TPR 195!
TANZANIA. Lushoto District: Mombo Forest Reserve, 22 Jun. 1962, *Semsei* 3488!; Mpanda District: Mahali Mts, Kasoje Village, 1 Apr. 1972, *Nishida* 121!; Morogoro District: Chazi, 21 Aug. 1951, *Greenway* 8627!
DISTR. U 2–4; K 3, 4, 7; T 1–4, 6–7; Z, P; Nigeria, Cameroon, Central African Republic, Congo (Kinshasa), Sudan, Ethiopia, Zambia, Mozambique, Zimbabwe, South Africa
HAB. Rainforest or swamp forest, riverine forest and secondary forest; 0–1500 m
USES. None recorded

SYN. *Motandra erlangeri* K. Schum. in E.J. 33: 318 (1903); Stapf in F.T.A. 4, 1: 613 (1902). Type: Ethiopia, Arusi, near Lake Awasa, *Ellenbeck* 1710 (B†, holo., K, lecto., chosen by de Kruif)
    Oncinotis oblanceolata Engl., nom. nud.; T.T.C.L.: 53 (1949)
    O. sp.; T.T.C.L.: 53 (1949)

INDEX TO APOCYNACEAE

111

No new names validated in this part

Milton Keynes UK
Ingram Content Group UK Ltd.
UKHW022109141024
449569UK00031B/1842